T0304057

Supply and Competition in Minor Metals

An interest in the minor metals – termed "minor" as their annual production is relatively small – had been developing for many years. This study, first published in 1965, examines patterns of supply that can be identified as underlying the production of minor metals, and then uses these patterns to investigate the nature and degree of competition in the production of minor metals. This book will be of interest to students of environmental studies.

Supply and Competition in Minor Metals

David B. Brooks

RFF PRESS
RESOURCES FOR THE FUTURE

First published in 1965
by Resources for the Future, Inc.

This edition first published in 2016 by Routledge
2 Park Square, Milton Park, Abingdon, Oxon, OX14 4RN
and by Routledge
711 Third Avenue, New York, NY 10017

Routledge is an imprint of the Taylor & Francis Group, an informa business

© 1965 Resources for the Future, Inc.

All rights reserved. No part of this book may be reprinted or reproduced or utilised in any form or by any electronic, mechanical, or other means, now known or hereafter invented, including photocopying and recording, or in any information storage or retrieval system, without permission in writing from the publishers.

Publisher's Note
The publisher has gone to great lengths to ensure the quality of this reprint but points out that some imperfections in the original copies may be apparent.

Disclaimer
The publisher has made every effort to trace copyright holders and welcomes correspondence from those they have been unable to contact.

A Library of Congress record exists under LC control number: 65026178

ISBN 13: 978-1-138-94182-3 (hbk)
ISBN 13: 978-1-315-67349-3 (ebk)

SUPPLY AND COMPETITION IN MINOR METALS

by David B. Brooks

RESOURCES FOR THE FUTURE, INC.
1755 Massachusetts Avenue, N.W., Washington, D.C. 20036

Distributed by THE JOHNS HOPKINS PRESS
Baltimore, Maryland 21218

© 1965, Resources for the Future, Inc.
Library of Congress Catalogue Card Number 65–26178

RESOURCES FOR THE FUTURE, INC.
1755 Massachusetts Avenue, N.W., Washington, D.C. 20036

Board of Directors: Reuben G. Gustavson, Chairman, Erwin D. Canham, Edward J. Cleary, Joseph L. Fisher, Luther H. Foster, Hugh L. Keenleyside, Edward S. Mason, Frank Pace, Jr., William S. Paley, Laurance S. Rockefeller, Stanley H. Ruttenberg, Lauren K. Soth, John W. Vanderwilt, P. F. Watzek. *Honorary Directors:* Horace M. Albright, Otto H. Liebers, Leslie A. Miller

President, Joseph L. Fisher
Vice-President, Irving K. Fox
Secretary-Treasurer, John E. Herbert

Resources for the Future is a non-profit corporation for research and education in the development, conservation, and use of natural resources. It was established in 1952 with the co-operation of the Ford Foundation and its activities since then have been financed by grants from that Foundation. Part of the work of Resources for the Future is carried out by its resident staff, part supported by grants to universities and other non-profit organizations. Unless otherwise stated, interpretations and conclusions in RFF publications are those of the authors; the organization takes responsibility for the selection of significant subjects for study, the competence of the researchers, and their freedom of inquiry. David B. Brooks is an RFF research associate in the Resource Appraisal Program. The illustrations were drawn by Federal Graphics.

Director of RFF publications, Henry Jarrett; *editor,* Vera W. Dodds; *associate editor,* Nora E. Roots.

FOREWORD

SO FAR as I know, this is the first attempt of this size and quality on the part of an economist—albeit one with several years of training as a geologist —to deal descriptively and analytically with the minor metals. Usually, these metals are lumped together in the last section or paragraph of a book on mineral economics, in a way that does little more than demonstrate the author's awareness of their existence.

The materials in this group are so diverse that an attempt to treat them systematically can lead one into useless or even tortuous generalizations. To refrain from categorizing, however, tends to keep the research in the sphere of the purely descriptive and means avoiding the challenge of analysis. In this report, David Brooks has tried to be sufficiently systematic to create some order in a previously unordered field without depriving the reader of enough descriptive material to permit him to follow the argument.

HANS H. LANDSBERG, *Director*
Resource Appraisal Program
Resources for the Future, Inc.

September 1965

PREFACE AND
ACKNOWLEDGMENTS

INTEREST in the minor metals—termed "minor" only because their annual production is relatively small—has been developing for many years. Some of the minor metals, such as arsenic, mercury (quicksilver), and platinum, have been known and applied to a multitude of nefarious and benevolent purposes since early historic or prehistoric time. A few others attracted some attention during the initial phases of the industrial revolution, but real interest in the minor metals did not come until the twentieth century. The situation changed moderately at first. Some metals, notably aluminum and the more important ferroalloys, moved up into the major metals group. There was interest in thoria (thorium dioxide) for gas mantles until the 1920's; in cesium as a coating on tungsten filaments in the late twenties; and in beryllium for hard copper alloys from the early thirties on. But by and large the limited availability and high cost of utilization, as well as lack of knowledge about their properties, kept minor metals well out of the mainstream of economic and technologic developments until World War II.

During the war years the environment for minor metals changed. Highly specialized requirements in modern aircraft, weaponry, and communications created a substantial market for minor metals. Indeed, they created the demand for properties that brought a number of minor metals into prominence. Older minor metals, such as bismuth and tungsten, were found useful in many new and quite different applications so that interest in them broadened. And a large number of metals, such as germanium and hafnium, that had been ignored or discarded in industrial wastes, were also found to possess properties with great advantages in modern applications. By the middle fifties minor metals were in the forefront of technologic development in the United States.

The rapidity of their rise from obscurity and their use in the growing aerospace and atomic industries led to an almost boundless optimism in the technical and profit potentials of minor metals. The environment had been set by uranium, which had risen from minor metal to major metal status over a period of just a few years. To many of the minor metals there were now applied more striking adjectives: the "new metals," the "strategic metals," the "mystery metals," and even the "wonder metals." The mining press in particular was quick to see in the minor metals new vistas of prosperity for the mining industry. (Nowhere is the optimism better expressed, nor the absence of the simplest principles of economics more evident, than in the following statement, which was quoted in *Engineering and Mining Journal*, November, 1955, p. 156: "There has been no previous mining of selenium, so there is no estimate as to what the [new] deposit might eventually be worth. But since the demand is so great, I doubt that price would be considered in Government purchases.")

The numerous warnings that the limitations of minor metals (again, both technical and economic) were severe went largely unheeded for some time. When the limitations could no longer be ignored, the adjectives changed again. Some minor metals came to be called the "military metals," indicating their limited applications and the absence of civilian markets. Most colorfully they were called "cinderella metals," when budget changes or recognition of the costs involved broke the spell. The case of titanium is best known. At one point it was thought that titanium would be specified as the construction metal for the next generation of military aircraft. Production of titanium metal (sponge) soared from 7,000 tons in 1955 to 17,000 tons in 1957. But missiles, for which weight is less critical, began to replace manned aircraft and within just two years titanium production had plummeted to less than 4,000 tons. (Even in November of 1957 *Business Week* was able to comment that "the producers of titanium, the wonder metal, are wondering where the wonder went.") Other metals underwent similar if less spectacular changes of fortune.

The successive booms and collapses stimulated sober re-evaluation of minor metals. For the most part both the enthusiasm of the middle fifties and the disillusionment with minor metals that followed have been dissipated. Each element does have a unique set of properties that makes it potentially useful in certain applications. However, physical performance alone does not determine which elements will actually be used; cost of utilization must also be considered. It is my hope that this study will lead to a better understanding of the economics of minor metals—so that they will less often be thought of as wonder metals and thus will not later have to be called cinderella metals.

This study first took form as a doctoral dissertation prepared for the University of Colorado, while I was a research intern at Resources for the Future. It is impossible to list all those—colleagues, professors, industry officials, and government personnel—who generously contributed both information and suggestions over the course of the study. However, special mention must be made of several persons. Morris E. Garnsey and Wesley C. Yordan of the University of Colorado read and commented upon various drafts of the manuscript while it was at the dissertation stage. Orris C. Herfindahl of Resources for the Future suggested the project to me, and then exercised day-to-day supervision of the dissertation with a high level of professional criticism. Later, Hans H. Landsberg, also of RFF, was indispensable in helping me broaden the study and convert the dissertation to its present form. In addition, Morris A. Adelman of the Massachusetts Institute of Technology and Bruce C. Netschert of National Economic Research Associates made extended comments on an earlier draft. Those who helped me bear no responsibility for the conclusions of the study, nor do they even agree with all of them. Such responsibility must remain my own.

DAVID B. BROOKS

CONTENTS

APPENDICES

LIST OF TABLES

LIST OF ILLUSTRATIONS

SUPPLY AND COMPETITION
IN MINOR METALS

1

INTRODUCTION

THE METALS industry has traditionally divided metals into two groups, major metals and minor metals. Major metals include the ten or so metallic elements—iron, copper, lead, zinc, gold, silver, etc.—that account for the great bulk of metal production and form the backbone of an industrial civilization. Minor metals are those metallic elements that are beyond the laboratory stage but are recovered each year in less than major metal tonnage or value; they are industrial commodities produced on a continuing basis, not just metals recovered for experimental or noncommercial purposes.

The number of major metals has not changed much over the past half century. With some exceptions the metals that were most important 50 years ago are most important today. Not so with minor metals. A list of minor metals compiled in 1920 would include few of the thirty to thirty-five elements which must be considered as minor metals today.[1] This increase in the number of minor metals is a reflection of the research effort that has been devoted to developing new and improved materials to satisfy the demands of modern technology. Demands for materials that have thermoelectric properties or are semiconductors, for metals with high (or low) nuclear cross-sections, or for structural materials with high strength-to-weight ratios are becoming commonplace in many fields. Indeed, although modern economies are frequently characterized by the vast quantities of materials they consume, they can equally well be characterized by the great variety of materials consumed. Minor metals, and the alloys and compounds of minor metals, are among the many materials to which we have turned to satisfy these new demands.

The elements that are treated as minor metals in this study are listed together with their chemical symbols just below. Note that two groups of elements are usually treated as if each were a single element. First, lanthanum and the fourteen elements following it in the Periodic System of the Elements

---

[1] G. A. Roush compiled a table showing the gradually increasing number of minerals and metals considered essential for national security in the interwar period, as shown by official and unofficial lists. *Strategic Mineral Supplies* (New York: McGraw-Hill Book Co., 1939), pp. 8–9.

are called collectively the rare-earth metals. Second, the five metals that are similar to platinum and occur with it are called the platinum-group metals.

Minor Metals

Antimony (Sb)	Hafnium (Hf)	Selenium (Se)
Arsenic (As)	Indium (In)	Sodium (Na)
Beryllium (Be)	Lithium (Li)	Tantalum (Ta)
Bismuth (Bi)	Mercury (Hg)	Tellurium (Te)
Cadmium (Cd)	Molybdenum (Mo)	Thallium (Tl)
Calcium (Ca)	Platinum-group metals†	Thorium (Th)
Cesium (Cs)	Radium (Ra)	Titanium (Ti)
Cobalt (Co)	Rare-earth metals (RE)‡	Tungsten (W)
Columbium (Cb)*	Rhenium (Re)	Vanadium (V)
Gallium (Ga)	Rubidium (Rb)	Yttrium (Y)
Germanium (Ge)	Scandium (Sc)	Zirconium (Zr)

* This element is referred to abroad and in chemical literature as niobium (Nb).

† Includes ruthenium, rhodium, palladium, osmium, iridium, and platinum. Platinum and palladium are the most common members of the group.

‡ Includes the following fifteen elements: lanthanum, cerium, praseodymium, neodymium, promethium, samarium, europium, gadolinium, terbium, dysprosium, holmium, erbium, thulium, ytterbium, and lutetium. Lanthanum, cerium, and neodymium are the most common members of the group.

Tables 1 and 2 give some idea of the range of demands that minor metals satisfied and the range of functions that they filled in the decade from 1955 to 1965.

Goals of the Study

Aside from their metallic nature—and that is not always apparent—minor metals seem to have little in common. Their properties certainly vary: lithium will float, but thallium and bismuth are among the heaviest natural elements; cesium, gallium, and mercury are liquids near room temperature, but rhenium and tungsten have the highest melting points of all metals; and so on. Even their production rates vary greatly. Annual world production of antimony, for example, is measured in tens of thousands of tons per year, whereas that of rhenium is measured in tens of pounds.[2] (Figure 1.) Prices

[2] Unless otherwise indicated, all production and consumption data are given in terms of the content of elemental metal (rather than gross weight) regardless of whether the data refer to source materials or to primary products. Tons always means short tons of 2,000 pounds.

range from a few cents per pound for white arsenic (As_2O_3) to more than a thousand dollars per pound for platinum, scandium, and others. Some minor metals, such as cadmium and germanium, occur as tiny fractions in a major metal ore and can be recovered only when large quantities of residual "waste" are accumulated at smelters or refineries. Mercury, in contrast, is processed into pure metal right at the mine site. Still other minor metals are recovered from minerals such as zircon that are themselves of considerable value. Some minor metals are established commodities with uses in a number of different industries; others have yet to find more than a single commercial application.

Despite this many-faceted diversity, it is often convenient to treat minor metals as a group. They are regarded as a group by industry and written about as a group in trade journals and textbooks. Joint supply is a common factor in their production, and intersubstitution in their consumption. In this study an attempt is made to determine whether further generalizations can be made. Thus, the underlying question is: In what sense, or to what degree, *can* minor metals be treated as a group?

To be more specific, this study focuses on the economics of minor metals, treating them as industrial commodities related by a mineral origin and by a relatively small annual rate of production. In recent years we have learned a great deal about the chemistry and physics and the engineering properties of minor metals, but we have learned far less about their "economic properties." Neither the economic relationships among minor metals nor the relationships of minor metals to other commodities have been studied to any extent. Commodity surveys published by the U.S. Bureau of Mines compile some economic data for each minor metal, as well as information on its sources, technology, and history.[3] But as commodity surveys they are deliberately short on analysis. For the most part, economic information on minor metals must be abstracted from the reports of a few other government agencies,[4] from the numerous journals going to management and professional personnel in industry, and from the occasional releases of the producing firms or an investment house interested in one of those firms.

[3] The most readily available of the Bureau of Mines commodity surveys are the chapter-length reviews of each metal that have been assembled into the *Mineral Facts and Problems* bulletins published every five years or so. Some minor metals have been covered in the current *Materials Survey* series of Information Circulars. When available they present considerably more information than is available in the bulletins.

[4] The reports of the Attorney General pursuant to Section 708(e) of the Defense Production Act of 1950 (printed by the Senate Committee on Banking and Currency) emphasize the effect of the DPA on competition and are most useful for the few minor metals covered. Reports of the Materials Advisory Board of the National Research Council treat minor metals from the point of view of national security, and frequently carry information on resource availability and production economics.

Table 2. Properties of Minor Metals Which Are Most Important in Their Metallic Uses

M—Property is that of metal itself; A--Property is that of an alloy or intermetallic; C—Property is that of a compound of the metal; X—Metal has poor or opposite characteristic for the given property

	Low weight	Melting point High	Melting point Low	High thermal conductivity	Elec. conductivity High	Elec. conductivity Low	Semiconductive	Thermoelectric	Other electrical properties	Good structural properties[1]	Ductile or easily worked	Hard or hardener	Low friction coefficient	Reactive	Deoxidizer or reductant	Corrosion resistant	Coloring agent	Poisonous	Low neutron cross section	Radioactive	Other
Antimony			A					A	A			A	A			A	C				C–fire resistant
Arsenic			M						A	X	A	A			A	A	C	C			C–decolorizer
Beryllium	M	MC		MA	A	C				MA	X	A				A			MC	M	A–nonsparking
Bismuth	X		A	X				A	A		A										
Cadmium			MA		A					A			M			M	C	MC	X		
Calcium										X					M	A					C–hygroscopic
Cesium								M	MC						M					M	
Cobalt		A										A					C			M	A–magnetic
Columbium		MA								M	A	A	A			A			A	M	
Gallium			MA		A	A			A			A				A					
Germanium			M				M		M			A				A					
Hafnium		M														M			X		
Indium			A				M		MA			A	M			A	C				
Lithium	MC		AC						A				M	A	A					M	C–hygroscopic
Mercury	X		M	X	M							X						MC			M–luminescent[2]
Molybdenum		MA		M						XA	AC					XA					M–low thermal expansion
Platinum Gr.	X	MA		MA						M			X			MA			M		
Radium																				M	M–n emitter
Rare-earths		MA			VARIABLE					A	MA	A			A	A			X		A–sparking, M–emit colors
Rhenium	X	MA							M		A	A				A					A–incandescent
Rubidium									M						M					M	M–absorbs light
Scandium	A	A																			
Selenium							M		M		A	A				A	C	C			
Sodium															M	A					M–luminescent[2]
Tantalum		MA						M		M	A	M	C	X		MA			M		
Tellurium							AC			A	A	A				A	C				
Thallium	X						M		M								C				M–luminescent[2]
Thorium	X	MC								A										M	M–B emitter
Titanium	M	A		X	M					MA	X	C	X		A	MA	C				
Tungsten	X	MA								XA	AC					MA					M–incandescent
Vanadium		MA							A	MA	A			A	A			M			A–magnetic
Yttrium		M						C			A							M			C–transmits microwaves
Zirconium		C								M					A	M			M		A–magnetic

[1] Strong and stiff relative to weight, creep resistance, fatigue strength, etc.   [2] As a vapor.

Table 1. Principal Uses of the Minor Metals and of Their Alloys and Compounds

M—used as a metal; C—used as a compound; A—used as an alloy; Letters in ( ) indicate major use of the element

	Ferrous alloys (steel making)	Electrical industry	Electronics industry	Chemical & petroleum	Food & textile industries	Glass, ceramics & refractories	Paints, pigments & enamels	Structural metal: aircraft	Structural metal: other	Nuclear reactors	Storage batteries	Bearing metal	Soldering, welding, & brazing	Machine tools	Control & measuring equipment	Decorative & protective uses	Munitions & explosives	Insecticides & fungicides	Drugs & pharmaceuticals	Other
Antimony				C	(C)	C					(A)	(A)	A		A	A	AC			
Arsenic			C			C											M	(C)	C	
Beryllium						(C)		M	A					(A)	M					
Bismuth			C			C			M				A		(A)				(C)	
Cadmium			A	C		C		M	M	(M)	A	A				(M)				
Calcium	M		M						A			A								Reductant
Cesium			M	M		(C)									M					Fuel
Cobalt	(M)					C							(A)							Magnets, alloys[1]
Columbium	(M)		M										A	A						
Gallium			A													(M)				
Germanium		A	(M)			C														
Hafnium										(M)										
Indium		A(M)				C					(A)	M								
Lithium			M(C)			(C)							C							Lubricants
Mercury		M						M			M				(M)			M	M	
Molybdenum	(M)	M		C					A											Alloys[1]
Platinum Gr.		MA		(M)											A	(MA)				Catalyst
Radium															C					Medical
Rare-earths	M		M	C		M		A	A											Carbon arcs
Rhenium			M												A	M				
Rubidium			M	MC		(C)	(C)												C	
Scandium																				Alloys[1]
Selenium		(M)	M	MC		MC	C												M	
Sodium				(M)						M										Reductant
Tantalum	(M)	M	MA			A								M	(M)	M				
Tellurium	(M)		C	MC		C	C											C		Alloys[1]
Thallium			C			C									C			(C)	C	
Thorium		C		C				(A)		(M)										Gas mantles
Titanium	M		MA				(C)	(M)	MA					C						
Tungsten	(M)	M	M	C		C								M	(C)					Alloys[1]
Vanadium	(M)			C		C				M										Catalysts
Yttrium			C																	Nodular iron
Zirconium	M	M	M		M	(C)				M					M	M				

[1] See properties in Table 2.

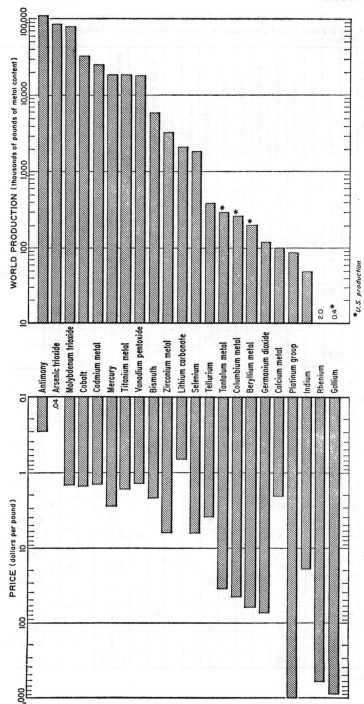

Figure 1. World production and prices of selected minor metal products, 1960. (U.S. production is shown for a metal when a world production figure was not available.)

The various surveys, reports, and journal articles provide a substantial and useful background of information on the economics of minor metals. However, the information is widely dispersed and typically is unemphasized in presentation because it is ancillary to another issue. Thus, a main objective of this study is to assemble the available information on minor metals and show how it can be used to analyze certain characteristics of minor metals or of groups of minor metals.

But a description of the supply and production of minor metals is not enough. Are there economic uniformities that will help answer questions concerning the relationships of supply to demand? For which minor metals do high prices and limited production in the recent past indicate continued scarcity? For which do the resources permit expanded production at constant or even declining unit costs? What role has technologic development played? Can we draw any conclusions from the cost-price record of recent decades when the demand for minor metals has been growing strongly? On the answers to such questions rest, at least in part, the appropriate decisions for both public and private policies involving minor metals. Some of the questions are beyond the scope of this study, and perhaps beyond the scope of available information. Moreover, the great range in the methods of producing minor metals, and the even greater range of applications for them, warn us from expecting broad or unqualified generalizations. For some purposes and some stages of production it does make sense to regard the minor metals as a group; for other purposes and other stages "the minor metals" are nothing more than a miscellany of metallic commodities whose lack of uniformity defies useful generalization. Nevertheless, by investigating the economics of minor metals as a group, we can provide a better framework for both the questions and the answers.

A Brief Review of the Study

The two-part organization of the book follows from the difficulties of treating a collection of diverse commodities. Part I deals with patterns of supply that can be identified as underlying the production of minor metals. It provides a framework for Part II, which uses these supply patterns to investigate the nature and degree of competition in the production of minor metals.

Part I: The Supply of Minor Metals

Chapter 2 outlines the important differences between the supply and demand conditions for minor metals and those for major metals. In contrast to major metals, the demand for many minor metals has been limited to a few markets and subject to rapid substitution. As a result, the demand side of economic analysis is not a promising vantage point from which to describe the minor metals. The supply side is more fruitful. Minor metals differ from major metals in that most of them are mined as joint products and do not become independent economic goods until later stages of production. The supply conditions of minor metals are related primarily to the forms taken by joint production, which in turn are based on the geologic conditions of occurrence and the stage of production at which minor metals become independent products. Both of these factors rest on physical relationships that can be regarded as constant and that can be ordered for purposes of analysis. In short, supply conditions explain much of the economic behavior of minor metals, and therefore organization of the minor metals on the basis of supply conditions is a useful approach to an investigation of production and competition.

In Chapter 3 the varying forms that joint production takes are used to classify minor metals into four main groups. And in Chapter 4 each of these classes is associated with a set of supply conditions that is distinct from those of other classes and that explains much of the economic behavior of the metals within the class. Byproducts extracted from a major metal during metallurgical processing are characterized both by the most restricted raw material supply conditions and by the least direct relationship between their own price and production rate. Individually mined minor metals exhibit a direct relationship between price and production until diminishing returns set in. Byproducts and coproducts of minor metals extracted during the milling of ores are intermediate in these respects. Finally, there are a few minor metals without effective geologic restrictions on supply, and for these metals the usual problems associated with mineral resources are unimportant.

Part II: The Production of Minor Metals

Part II is directed toward evaluating the strength of competition in the output of minor metals. It is essential to evaluate competition in order to understand the economics of minor metals on the basis of price and cost data. Strong competition in an industry will lead to a situation in which profits greater than those necessary to attract capital tend to be eliminated by increases in production capacity and, conversely, losses tend to be eliminated by decreases in production capacity. When this is the case, prices will reflect costs, and can be used to measure changes in the economic efficiency of production, which in turn will reflect the effects of changes in resource quality and in technologic efficiency.[5]

[5] Orris C. Herfindahl, *Copper Costs and Prices: 1870–1957* (Baltimore: The Johns Hopkins Press for Resources for the Future, 1959), pp. 1–3.

The strength of competition in the production of minor metals can be evaluated by comparing the actual price-quantity behavior of minor metals in recent years with the behavior that would be expected according to the classification developed in Part I. This is done in Chapter 8, the concluding chapter of the study. However, consideration of the strength of competition takes on more meaning if we also have information about its nature and sources, and if we can relate these aspects as well to the supply conditions. This picture is built up in Chapters 5, 6, and 7, which describe the structure of minor metals production.

In Chapter 5 the location, scale, and rate of growth of minor metals production are investigated. It is shown that minor metals production cannot be regarded as a separate industry made up of firms that emphasize the mining or recovery of minor metals. Rather, it is characterized by the presence of firms that produce minor metals as but a small part of their overall activities. Most firms apparently specializing in the production of minor metals turn out on closer inspection to be affiliates of larger concerns. Except with titanium, joint ventures have not been typical of minor metals production, but various forms of technical and marketing agreements are common.

Questions about the structure of minor metals production must be carefully phrased, for there is no industry as ordinarily defined, and the usual questions may have no meaning applied to the firms involved. One cannot, for example, analyze the general policies of some firm nor find out why the firm is large just on the basis of questions about its role in minor metals production. But one can ask why that firm is producing minor metals along with its other activities. Specifically, one can ask such questions as why certain firms produce certain minor metals but not others, and why some firms produce just one minor metal whereas other firms produce several. These matters are discussed in Chapter 6, which deals with the firms producing minor metals. It is shown that firms from the nonferrous metals industry dominate the production of metallurgical byproducts of major metals and that firms from the chemical industry dominate the production of most minor metals in two other classes. The successful producers are characterized by large size, partial vertical integration, and diversification into production of several minor metals. Only the producers of beryllium and mercury tend to be specialized to one metal, the former for historic and the latter for geologic reasons.

The two final chapters deal more directly with competition. Chapter 7 emphasizes the sources of competition. First, the several factors that lead firms to enter into the production of minor metals are considered, and then the sources of competition among existing producers are examined. Chapter 8, as indicated above, reviews the postwar price record for minor metals in light of the information and the models developed in preceding chapters. In brief, the raw materials which firms control or handle *for purposes other than production of minor metals* have a great deal to do with determining which firms produce which minor metals. Other important forces associated with entry include the technologic similarity of the production processes for certain minor metals and the tendency for some manufacturing firms to integrate backward into primary production of the metals they need.

Technical economies of scale, low and unpredictable levels of demand, and the advantages of integrated production, combined with the inability of many firms to secure minor metal-bearing source materials, lead to a high degree of concentration in the production of minor metals. In the case of a few metals these factors are sufficient to permit monopolistic influence in the markets. For most minor metals, however, further opportunities for entry do exist, and, together with the possibilities for substitution, the desire of producers to expand markets, and the availability of imports, they over-balance the restrictive forces and create fairly competitive markets in the United States for these minor metals and for their source materials.

Some Final Comments

Aside from Chapter 4, which in part is addressed to the economist, no special knowledge of either economics or of minor metals technology is expected of the reader. Terms such as supply, demand, and economies of scale are taken from the economist's vocabulary but they can be readily understood in context. Technical terminology has been largely avoided. A few terms, notably *metal, source material,* and *primary product,* deserve more careful definition and are considered further in the next chapter. In some tables it has been convenient to use chemical symbols rather than the names of the elements. Appendix A presents a list of elements alphabetized by symbol for the reader unfamiliar with them. It has also been convenient to use the common names of certain corporations rather than their longer official names: Amax for American Metal Climax, Inc., and TMCA for Titanium Metals Corporation of America, for example. In most cases the corporation can be readily identified from the abbreviated name, but Appendix B provides an alphabetized list for cases in which confusion might arise.

Two additional appendices have to do with minor metal statistics. Because production and consumption data on even a few minor metals are rarely brought together in one place, it is difficult to compare them with one another or with other commodities.[6] In order to rectify this situation,

---

[6] Statistical data for many minor metals are published annually by the Bureau of Mines in *Minerals Yearbook,* Vol. I, *Minerals and Metals (Except Fuels.)* As explained in Appendix C these data are not necessarily consistent from metal to metal.

Appendix C presents a cross-sectional statistical picture of minor metals in the United States economy and of their place in world markets in 1960. It also includes a discussion of the problems entailed in collecting data on minor metals and putting them into comparable form and consistent units. Appendix D is a brief description of the techniques and the standards used to construct the indexes of minor metal production that are presented in Chapter 5.

In its more detailed aspects this study is limited to the economics of minor metals in the United States. The limitation is partly a matter of convenience, for data on production in other countries are neither readily obtainable nor comparable with those on the domestic industry. The limitation is also justified by the position of the United States in the world supply and demand picture. For some minor metals the United States is a major producer of source materials; for most it is a major producer of primary products; and for all of them the United States is a major consuming nation. However, the competitive influence of foreign production must still be taken into account even though it does not receive the detailed study given to domestic production.

PART I

THE SUPPLY
OF MINOR METALS

2

MINOR METALS
AND MAJOR METALS

WHY DISTINGUISH minor metals from other metals, or, for that matter, metals from other commodities? The answer in either instance is that they have somewhat different supply and demand conditions, and these differences are reflected in somewhat different economic behavior. The economic behavior of minor metals is, of course, influenced by many factors—the fact that they are producer goods, that they are capital goods, and so forth—but these other factors are common to many economic goods and can remain implicit in discussion. The emphasis in this chapter is on the characteristics of minor metals that distinguish them from major metals.

What Are "The Minor Metals"?

Discussions of minor metals are often confusing because they exclude elements that may not be metals by some chemical definition or because they include minerals that contain minor metals. The thirty-three metallic elements (or groups of elements) that are treated as minor metals in this study have been selected from the more than ninety naturally occcurring elements on the basis of two criteria: the first distinguishes metals from minerals, and the second minor metals from major metals. However, as will be apparent, the criteria are flexible. At some points it is instructive to mention magnesium and high-purity silicon although for the most part they are not regarded as minor metals.

Metals Versus Minerals

The once-clear distinction between metals and nonmetals has become increasingly vague with advances in solid state physics. Common definitions of *metal* in terms of physical properties, such as ductility and malleability, are all right so far as they go, but they really say only that a metal is a substance possessing metallic properties. Modern definitions in terms of electron bonding are also clear only for those elements which would be metals under any definition.

For the purpose of economic analysis a more useful distinction is that between metals and minerals. In nature metals are generally found contained in *minerals*, which are naturally occurring, inorganic chemical compounds. A *rock* is nothing more than a natural mixture of minerals which may or may not—usually the latter—be valuable. A metal is therefore a commodity which is *not* used as it occurs in nature, but which *must be extracted from a mineral and changed to a new form* before satisfying our demands for it. This proposed distinction between metals and minerals is based on the character of the production process rather than on any estimate of "metalness." It separates those minerals that are mined for the elements they contain from those minerals that are mined for the properties of the mineral as an entity. Thus not only typical metals but also the metalloids and semiconductors, which are left in limbo by older definitions, are included as metals. Certainly there is more in common, at least in some ways, between the production of calcium metal and cobalt, for example, than between calcium metal and limestone (natural calcium carbonate).

Minor Metals Versus Major Metals

The simplest way to separate minor metals from major metals is by either value or volume of annual world production.[1]

A maximum annual value for minor metals is difficult to fix because exchange rates vary and because there is no one point in the production processes for minor metals at which their values are all comparable. Fortunately, a value limitation is needed only for gold and silver, which are better considered as major metals than as minor metals because of the large and widespread industry devoted to their recovery.

A convenient dividing line between minor and major metals can be set at world production of about 75,000 tons per year in primary products. This line seems to separate most metals commonly regarded as major from those regarded as minor. However, world production of sodium (classified as minor) does exceed this limit, and that of uranium (classified as major) falls below it. (Also, as indicated above, those metals that have been isolated on no more than a laboratory scale are not regarded as minor metals.) The

[1] It is necessary to use world rather than domestic production figures so that metals such as tin and manganese, for which domestic production is negligible, will be properly defined as major metals.

limit places iron, copper, lead, zinc, tin, aluminum, and magnesium in the category of major metals along with gold, silver, and uranium, and it separates the three large tonnage ferroalloy metals (manganese, chromium, and nickel) from other ferroalloys.

The number of years that a metal has been in use is a poor guide to whether it is a major or a minor metal. Though a preponderance of the minor metals have been developed since 1940, some minor metals are at least as old as most major metals and some major metals are no older than most minor metals.

Supply

In contrast with major metals, the production processes for minor metals cannot as a rule be regarded as an extended input-output series separated by nothing more than time lags and incomplete recovery of metal in processing. Major metal sources, such as copper ore, are valuable only because of their metal content; the object of mining copper ore is to recover copper metal, and basically there is just a single series of processes from mine to refined and fabricated copper. This is not the case for the majority of minor metals. Most of them are originally found in joint sources with other valuable minerals or metals, and for many the primary product is not elemental metal. As a result there is a divergence between the supply conditions for source materials and the supply conditions for the primary products derived from them. Both aspects of minor metal supply require further attention, but the importance of the divergence is clearer when the production of minor metals is divided into three stages.

Stages of Production

The first stage of production is mining, which involves the physical removal of metalliferous rock from the earth and bringing it to the surface. The second stage is called ore dressing or milling. In milling the several *mineral* constituents of the ore are separated from one another and from waste rock. For products used as minerals, milling is commonly the final stage of production, but for metals it is just an intermediate stage. Valuable metal-bearing minerals are concentrated and sent on to the next stage, while waste "gangue" minerals go to the dump. All minor metals go through the stages of mining and milling, but some do not form distinct commodities until the third stage—metallurgical reduction and refining. In this stage, chemical or metallurgical techniques are utilized to decompose minerals into their constituent ions and to reduce metallic ions to primary forms.[2] Statistically speaking, there are two phases, consumption of source materials and production of primary products, related as input and output. The metallurgical stage comprises all processes through the commercial production and sale of the first industrial commodity in which the minor metal in question is the element of predominant value. It includes any intermediate steps in the course of producing a market product (such as the production of tetrachlorides, which are intermediate to the production of sponge metals by the Kroll process). But it does not include the refining of metals to more than commercial-grade purity or the melting or forming or semifabricating of metals, since these processes use primary products as input. For the same reason, it does not include the production of compounds or alloys other than those, such as antimonial lead, that are recovered directly from source material as alloys or compounds.

Source Materials

The three-stage scheme described above is followed by all of the minor metals. However, the variations within the scheme are many. Just a few minor metals, most notably mercury, are similar to major metals in that the primary products are recovered from ore mined with the sole intention of producing that metal. The term *ore* is universally reserved for mineral deposits that can be mined or quarried *at a profit* given current market conditions and available technology. Moreover, when the term *ore* is combined with the name of a metal, as in *tungsten ore*, the reference is to a mineral deposit that can be worked profitably and in which that metal is the product of main value. Therefore, those minor metals that are produced as by-products do not have an ore source under their own names, and they are typically not recorded in statistics for *mine production*.[3] The term *source material*

[2] Any text on metallurgy will describe the processes used. A brief summary of mining, milling, and metallurgical techniques is presented in *Economics of the Mineral Industries*, Chapter 1, "Distinctive Features of the Mineral Industries," by Charles H. Behre, Jr., and Nathaniel Arbiter. (New York: AIME, 1959).

The distinction between the processes to be included in milling and those to be included in the metallurgical stages is not always clear. Here the practice followed has been that of the Bureau of the Census, which distinguishes between Mineral Industries (mining and milling) and Primary Metals Industries (further processing). The former includes mines plus "all ore dressing and beneficiating operations. . . . These include mills which crush, grind, wash, dry, sinter, or leach ore, or perform gravity separation or flotation operations" (U.S. Bureau of the Budget, *Standard Industrial Classification Manual*, 1957; "Metal Mining," p. 22). The inclusion of leaching as a milling operation is open to question but has been retained.

[3] The Bureau of Mines records mine production for a metal so long as a distinct concentrate of minerals containing that metal as the main value is segregated at the mine or mill. The Bureau of the Census does not record mine production for a metal unless that metal is the main value produced in the entire mine; a mine is a copper mine if copper

is here used to refer to all commercial raw sources of a metal, including both ore and byproduct sources, regardless of whether the latter are "wastes" of some other process.

The majority of minor metals are recovered as joint products, either at the milling stage of production or at the metallurgical reduction and refining stage. In either instance, the presence of minor metals may have had little influence on the mining of the original source material.[4] The immediate minor metal source material is commonly the residual tailings, slags, drosses, and other "wastes" left after recovery of the major metal.

Whenever a metal is not recovered entirely from its own ore, that is, whenever joint production is an element in its supply, there will be a divergence in supply conditions between source material and primary product. For example, the output of primary product may exceed "mine production" if some of the metal in question is recovered as a byproduct. Or the amount of ore consumed may exceed the quantity of metal produced if some of the ore is used for other purposes. Among milling joint products the proportion of the byproduct that is used as a source of metal may be very small. For example, only 10 per cent of the zircon recovered is used as a source of zirconium metal and hafnium metal; the bulk of it is consumed in mineral form as ceramic or foundry sand.

Primary Products

The supply conditions of minor metals are just as complex for the primary products as for the source materials. The primary product is the output of the metallurgical reduction and refining stage of production; though it is generally called a "minor metal," it may or may not be elemental metal. Obviously, *metal* is being used in two ways. In some cases it refers to a metallic element in general or to metallic elements collectively. It is used in this sense in the term *minor metal*. In other cases *metal* refers to an element in its reduced (elemental) state, and in such cases it is combined with the name

---

contributes the largest share of the value of metals produced, regardless of the fact that a molybdenite byproduct may be recovered at the mill and a selenium byproduct at the refinery. Moreover, the classification of industries used by the Bureau of the Census divides the Mineral Industries from the Manufacturing Industries at the point between mine and metal production. However much sense this distinction makes in the case of major metals, it is unsatisfactory for the collection of minor metal statistics. Most minor metals do not show up at all in the Census statistics or are scattered through a variety of classification headings. In Appendix C an attempt has been made to estimate the mine production of certain metals in all source materials, rather than just those coming from ores.

[4] There are exceptions. The content of molybdenum-bearing minerals in low-grade copper ores can be recovered in milling operations and is often an important source of revenue to the copper mining firm. In contrast with most minor metals, byproduct gold and silver are always taken into account because of the ready market and their relatively high values per ton of ore.

of the element—*columbium metal*, for example, means commercially pure columbium, neither alloyed nor compounded with other elements.

If a *primary minor metal product*, or more simply a *primary product*, need not be elemental metal, how is it to be identified? Since the primary product is the output of the metallurgical reduction and refining stage of production, it must be the first form of a minor metal commercially traded or consumed after that metal has been recovered as a separate commodity from its source material. (In some cases the primary product consists of a pair or a group of chemically very similar metals that are produced and consumed as a unit, as with cesium-rubidium or the rare-earth metals.) The primary product thus lies between metal in a source material, on the one hand, and metallic elements that have undergone additional processing or scrap metal (sometimes called *secondary metal*), on the other.

Table 3 associates each of the minor metals included in the study with its important primary products. For most minor metals there is no trouble identifying the primary product. In general, there is some key stage of production at which nearly all of a given metal is in a single form, which can be taken as the primary product. It is either marketed in that form or used in that form as the input material for conversion to a variety of other forms. Almost all cadmium, for example, is processed directly into cadmium metal and sold in this form after being recovered from zinc plant flue dust. The bulk of molybdenum is first recovered from its sources as molybdenum trioxide (molybdic oxide), and it is from this one compound that molybdenum metal, ferromolybdenum, and molybdenum compounds are prepared. Other metals are represented by several primary products. Antimony, for example, has three; it is recovered from antimony ore as elemental metal and as a trioxide, and it is also recovered directly from impure lead ore as antimonial lead alloy.

In a few cases the choice of the primary minor metal product is an arbitrary one. Some might wish to consider titanium ingot (purified titanium) rather than sponge metal as the primary product. Others might suggest titanium tetrachloride. Decisions were based largely on the commodity that seemed to be regarded as the primary product in commerce. Often the choice is indicated by the fact that statistical data are collected for one form of the commodity, which is the case with sponge titanium, or by the fact that all data are reported in terms of a standard commodity, as with lithium, which is all reported as if it were recovered as lithium carbonate.

For a number of metals, a small sector of production, typically the direct manufacture of chemicals from source materials, has been omitted in order to limit the scope of the study. These products are listed in column 3 of Table 3. The discussion of the structure of minor metals production in later portions of the study does not necessarily refer to these less important sectors,

Table 3. Primary Minor Metal Products

Metal (1)	Important primary products (those included in the analysis and the data tables)[1] (2)	Less important primary products (those excluded from the analysis and data tables)[2] (3)
Antimony	Metal; Antimonial lead; Trioxide	
Arsenic	Arsenic trioxide (white arsenic)	
Beryllium	Metal	Oxide
	Copper alloy	Other alloys
Bismuth	Metal	Lead alloys
Cadmium	Metal	Compounds
Calcium	Metal	Not applicable
Cesium	Metal; Compounds	
Cobalt	Metal; Oxide	
Columbium	Metal; Ferrocolumbium	Compounds
Gallium	Metal	
Germanium	Metal or Dioxide	
Hafnium	Dioxide or Sponge metal	
Indium	Metal	Compounds
Lithium	Carbonate (all other products are reported as carbonate equivalents)	
Mercury	Metal	
Molybdenum	Molybdenum trioxide	"Moly" added directly to steel
Platinum-group metals	Platinum, Palladium, Iridium, Osmium, Rhodium, Ruthenium (as metals)	
Radium	Metal and compounds	
Rare-earth metals	Mixed compounds	Misch metal; Individ. metals
Rhenium	Metal; Ammonium perrhenate	
Rubidium	Metal; Compounds	
Scandium	Scandium oxide	
Selenium	Metal	
Sodium	Metal	Not applicable
Tantalum	Metal; Ferrotantalum-columbium	Compounds
Tellurium	Metal	
Thallium	Metal; Sulfate	
Thorium[3]	Nitrate; Metal (alloy-grade)	
Titanium	Sponge metal	Ferroalloys; Compounds
Tungsten	Tungsten trioxide	Other compounds, alloys and metal
Vanadium	Vanadium pentoxide	
Yttrium	Yttrium oxide	
Zirconium	Sponge metal (hafnium-free)	Ferroalloys and other alloys

[1] The commodities listed in this column are those considered to be the important metal products in terms of the criteria specified in the text. It should be emphasized that, in general, data on primary products in the statistical tables (text tables 5, 6, 9, and 15 and the tables in Appendix C) refer to the commodities listed in column 2, and only to these commodities.

[2] Commodities listed in this column are those products of the metallurgical reduction and refining stage that were excluded from the analysis: (1) because they were not important (as with primary cadmium compounds); or (2) because information on them was unavailable (as with bismuth-lead alloys); or (3) because information on them could be included with the commodity listed in column 2 (as with lithium compounds). In a few instances data for commodities listed in column 3 will be shown in the statistical tables, but in all such cases they will be clearly labeled.

[3] Only non-energy uses and production.

nor are they generally included in statistical tabulations. In three cases larger sectors of production have been excluded. With calcium and sodium, attention is restricted to the production of elemental metal in order to illustrate a particular resource supply situation, and with titanium the whole production sequence leading to titanium pigments has been excluded because of its size.

Demand

Throughout this study, the supply of minor metals is viewed apart from the demand for them. This does not mean that the demand side of economic analysis is ignored. Rather, it means that we assume its presence as an independent or exogenous force but do not find it necessary to analyze the factors underlying demand conditions or changes in demand conditions.

It is fortunate that demand can be assumed rather than analyzed, for the demand for minor metals is not susceptible to easy ordering. By and large, minor metals lack the broad, stable, or predictable markets that are characteristic of major metals. Demand conditions vary widely from metal to metal and from year to year. Among the reasons for this lack of stability is the nearly total dependence of certain minor metals on a single use. The development of a more efficient substitute can reduce or even eliminate the need for the minor metal in that use. The consumption of germanium, for instance, depends largely on a few electronic components in which it is subject to increasing substitution by high-purity silicon.[5] More efficient use of germanium itself has also cut the rate of consumption below what was expected a few years ago. Finally, consumption of the minor metal can shift because of changes in demand for the commodity in which it is used. Thoria (thorium dioxide) is still the most efficient material for gas mantles, but thorium consumption has never regained the levels of the first two decades of this century when gas lights were the principal means of illumination.

In some cases the problem of a limited number of applications is compounded because there is only one purchaser. The production of zirconium sponge metal, for example, has rested almost entirely on AEC production contracts, and there is no certainty as to what the market situation will be after these contracts expire. Only somewhat less vulnerable are those minor metals, or particular primary forms of minor metals, whose consumption remains tied exclusively to the aerospace or defense industries.[6] Efforts by

[5] For an interesting discussion, see "Sister Metals Squabble for Markets," *Chemical Week*, vol. 91 (December 8, 1962), pp. 73–79.

[6] *American Metal Market*, November 27, 1964, p. 20, and January 14, 1965, p. 10: see also the Special Sections on the newer metals, July 23, 1963, 22 pp; and June 22, 1964, 16. pp.

producers to broaden the consumption base of the numerous minor metal products that fit into this category have not been uniformly successful.[7]

Finally, it must be remembered that plastics, ceramics, and metal-nonmetal composites are capable of replacing metals in many old applications and restricting their use in new ones.[8] Since the relevant price to the buyer is the total cost of utilization, including costs of fabrication, joining, and so forth, not just the cost of the material itself, many nonmetallic materials compare very favorably with minor metals when these additional costs are considered. If a more expensive material is selected, the buyer must be able to justify the cost both in terms of the added performance obtainable and in terms of the need for that added performance in the final product.

Some minor metals do have specialized markets in which they are reasonably secure. Bismuth is not likely to be rapidly replaced in pharmaceuticals, nor is antimonial lead in storage battery plates. Even some of the newer minor metals, such as beryllium in copper alloy, have markets in which there are no important substitutes as yet. However, consumption in the secure sector must be a fair proportion of total consumption if the stabilizing effect is to be important. The fact that titanium is assured a market in certain corrosion-resistant chemical processing equipment is not sufficient to protect it from fluctuations in consumption, for this use accounts for only about 5 per cent of annual titanium metal consumption, whereas military aircraft and missiles account for 75 per cent.

The record in recent years suggests that for many minor metals the responsiveness of consumption to price changes is much less in the short run than in the long run. Price elasticity of demand is somewhat lower in the short run than in the long for nearly all commodities, as many factors prevent a manufacturer from shifting quickly from one commodity to another as relative prices change. But with minor metals, and particularly with the newer minor metals, the difference may be extreme.

The short-run inelasticity of demand can be seen in the histories of several minor metals. Titanium and cobalt producers, for example, expected great increases in consumption when they lowered the price of primary metal. In neither case was this the result, and both industries have now shifted tactics. Price is being held relatively constant, and the industries are supporting

[7] *American Metal Market*, June 22, 1964, Section II, p. 10. *Wall Street Journal*, March 2, 1962, p. 1. Many articles on individual metals highlight this same problem.

[8] On competition between minor metals and plastics, see *American Metal Market*, January 17, 1962, p. 1; *Barron's*, "Plastics vs. Metals," Vol. 41 (September 4, 1961), pp. 11–12; *The Journal of Commerce*, June 29, 1962, p. 6, and August 31, 1964, p. 5A. On ceramics and cermets, see Robert J. Fabian, "Aerospace Materials: Where We Stand Today," *Materials in Design Engineering* (December 1961), pp. 94–97; *American Metal Market*, February 25, 1963, p. 13, and December 28, 1962, p. 10. On composites of fibers and plastics with metals, see *American Metal Market*, December 12, 1961, p. 14.

research studies in the hope of finding new uses for their products. Formation of such groups as the Selenium-Tellurium Development Committee and the Zirconium Association is indicative of the same kind of effort for other metals.

If events of recent years have proved anything, it is that there is no fixed demand for major metals, much less for minor metals; there is only a demand for materials that can perform certain functions in certain ranges of cost. Given this situation, it may be that the concept of long-run demand does not have much meaning for many minor metals, certainly not for the newer metals. Rather, there is a series of new and sometimes quite different short-run demand situations. As a result, it is not possible to take a given demand situation and project to the period when expansion schedules will be complete and other dynamic forces will have worked themselves out.

Rarity and Scarcity

Some minor metals are characterized by physical *rarity*, some by economic *scarcity*, and some by both. There is no necessary relationship between rarity and scarcity. A scarce metal may also be rare, but it need not be; a rare metal is commonly also scarce, but it may not be.

Rarity is determined by an element's relative physical abundance in some specified portion of the earth, usually the crustal zone (lithosphere) with or without the lakes and oceans. Thus, the rarity of any element is a constant (neglecting radioactive decay) and can be estimated by a number of techniques with more or less accuracy.[9] It is ordinarily expressed in percentages or in parts per million (ppm) by weight. Iron, which makes up 50,000 ppm (5 per cent) of the crust is less rare than zinc at 25 ppm, which in turn is less rare than gold at 0.005 ppm. Scarcity, in contrast, is determined by cost of acquisition under given conditions of time and place. Thus, the scarcity of any element must be expressed in value terms, rather than in physical terms, and it is by no means a constant. Zinc is ordinarily more valuable than manganese, but under wartime scarcity conditions, manganese may become equally valuable.

The terms rarity and scarcity not only have different meanings but they also derive from different origins. Rarity derives from the geochemical differentiation of the universe, a force over which man has no control. Scarcity can result from many causes, among which are some, such as monopoly, over which man has a great deal of control. However, there are

[9] Brian Mason, *Principles of Geochemistry* (2nd ed; New York: John Wiley and Sons, Inc. 1958), pp. 41–43. The most recent compilation of data on the abundance of elements in the earth's crust has been prepared by A. P. Vinogradov, "Average Contents of Chemical Elements in the Principal Types of Igneous Rocks of the Earth's Crust," *Geochemistry* (English translation), No. 7 (1962), pp. 641–52.

three fundamental geochemical reasons why an element may be scarce—though the effect of any of the three can be mitigated by developments in technology.

First, an element may be scarce because it is rare. Radium, rhenium, tellurium, and the platinum-group metals are among the rarest elements. But this resource base scarcity can only be a very partial answer. Titanium is one of the nine most abundant elements; rubidium is six times as abundant as copper and more than 20 times as abundant as lead; lead itself is about equal in abundance to scandium. Yet in each case there currently exist sharp differences in scarcity, differences which are obviously not accounted for by rarity.

Second, and more important than abundance in determining scarcity, is the amount by which a metal in an ore deposit is concentrated over its *average* crustal abundance.[10] Some metals are not concentrated by a very large factor into ore deposits. In its major source rubidium is concentrated barely five times, whereas a one per cent copper deposit has been concentrated 180 times, and a two per cent lead deposit more than 1,300 times over their respective average crustal abundances. In other words, workable deposits of major metals both in terms of size and of grade are more common than are those of most minor metals. The question of why this is so is one of the main problems of geochemistry.

The third geochemical reason for scarcity comes from the problem of extracting metals from minerals and rocks. Many metals are expensive because they cannot be won from the rock masses which contain them, or more commonly from sister metals of very similar chemical properties, without great difficulty. The necessity of separating most beryl, the main ore of beryllium, from associated minerals by hand is an example of the former problem and the pair columbium-tantalum (often referred to in just that way) of the latter.[11] Whenever it is difficult to separate metals from one another, the tendency is to use them in combined form as much as possible. For most ferroalloy uses the tantalum impurity in columbium has no ill effect and is not extracted (but for most uses of columbium metal, tantalum content must be reduced to a low level). Similarly, the rare-earth elements[12] are most commonly found naturally associated in a mineral, monazite, from

[10] Mason, *op. cit.*, pp. 47–48.

[11] Sister metals are almost invariably associated in nature. The other pairs of sister minor metals are zirconium-hafnium and cesium-rubidium. Also commonly associated are the metals selenium and tellurium, thorium and the lighter rare-earth metals (lanthanum through europium), and yttrium and the heavier rare-earth metals (gadolinium through lutetium).

[12] The use of *rare* in *rare-earth metals* is not dictated by the concept of abundance, for these elements are moderate in abundance, but is of historic origin and refers to the difficulty of separating them from one another. The Bureau of Mines has suggested that the term be used in the hyphenated adjective form to avoid confusion with truly rare metals.

which combined compounds can be easily extracted. But separation of individual rare-earth metals from one another is a process that has only recently become commercial and that still requires very careful, almost laboratory-like techniques. Thus, as much as possible, the rare-earth metals are used in combined form and are not separated into the individual elements.

The question of why minor metals are minor is, therefore, a many-faceted one in which rarity plays a part, but only a part. In addition to the aspects discussed, difficulties in fabrication and other factors operative at later stages of production give further answer to the question. Finally, there is one apparent aspect of scarcity that ought to be noted. It has often been stated that many metals are scarce because they have not been looked for. The case of uranium is commonly cited to "prove" that abundant, low-cost supplies of minor metals may exist unrecognized. Nevertheless, it is doubtful that any concentrations of minor metals comparable to those of the major metals have been overlooked. Their occurrence, rather, is different both qualitatively and quantitatively from that of major metals. It must not be forgotten that those abundant supplies of uranium have an average content of only a few tenths of one per cent uranium, a concentration well below the average grade of copper being mined in the United States today after a century of exploitation.

At the same time, there is something hopeful in the unusual nature and occurrence of minor metals. Discoveries may result from the use of new exploration techniques. The berylometer, for example, which utilizes certain unique atomic properties of beryllium, was instrumental in the discovery of the disseminated deposits of beryl in western United States. The "ores" may turn out to be not rock but brines, or the ashes of certain coals. The resource picture for columbium was changed not by finding more deposits of columbite, the traditional source, but by bringing deposits of pyrochlore, a new ore source, into production. Even in the United States and Western Europe, the areas best geologically explored in the world, the opportunities for finding new sources of minor metals are by no means exhausted.

3

A CLASSIFICATION
OF MINOR METALS

AS EMPHASIZED in the preceding chapter, where supply and demand conditions for minor metals were outlined in quite general terms, the most important ways in which minor metals differ from major metals, and the most important ways in which they differ from one another, relate to the forms taken by joint production. In this chapter a classification is developed which isolates these forms and which serves later as the basis for the construction of several models of the economic behavior of minor metals.

Organization of the Classification

The classification is really an attempt to sort out minor metals according to the dominant source material from which they are currently produced. It rests on two not entirely independent bases. Both are defined in physical terms, but refer to those characteristics of source materials that are the principal determinants of their supply conditions.

The first breakdown relates to the stage of production at which a minor metal (or a minor metal-bearing material) initially becomes a separate product, the stage, that is, where costs stemming from production of the metal become specifically attributable to it. This may be the metallurgical reduction and refining stage, the milling stage, or the mining stage.

The second breakdown indicates the joint product relationship (if any) of the metal at the stage when it becomes a separate product. Metals are distinguished as individual products or as coproducts or byproducts. An individual product is a metal which is produced with no associated metals or with associates of comparatively insignificant value. Metals that are separate products at the mining stage are by definition individual products. Metals that do not become separate products until later stages are byproducts or coproducts of a main product, which is usually one of the major metals.

The terms "byproduct" and "coproduct" are used to indicate the degree of influence of the minor metal on the production of the main product with which it is geologically associated. A metal is a byproduct if a change in its price has almost no influence on the output of the main product, and it is a coproduct if a change in its price has an appreciable influence on the output of associated coproducts.

Whether a commodity is a byproduct or a coproduct depends as much upon relative prices as it does upon its content in the source material. For example, thorium and rare-earth metals occur together in the mineral monazite. In the early decades of this century thorium was extracted from monazite for use in gas mantles, and the residues bearing rare-earth elements were discarded. With the decline in use of gas mantles and the rise in use of a rare-earth alloy, misch metal, to make flints, rare-earth metals became the main product of monazite and thorium the byproduct. The situation reversed itself once again with recognition of the importance of thorium in atomic energy. Today, both products are about equally important. In any event, joint product relationships can be thought of as ranges on a scale extending from a commodity of so little relative value that it exerts no influence on the production of the main product to a commodity which is mined as the sole product from its own ore.

Each byproduct or coproduct is further distinguished according to whether it is associated with a major metal or a minor metal. A minor metal can be a byproduct or a coproduct of another minor metal, but for all practical purposes it can be only a byproduct of a major metal. Finally, there are some metals that are extracted from natural sources of almost unlimited size. Because production from a source of this type is economically feasible for a few minor metals, and because it involves supply conditions different from those of depletable sources, a separate Class D has been defined even though it is not co-ordinate with the other classes. Thus, using both bases, the classification is as follows:

   A. Joint Products at the Metallurgical Reduction and Refining Stage
      1) Byproducts of Major Metals
      2) Byproducts and Coproducts of Minor Metals
   B. Joint products at the Milling Stage
      1) Byproducts of Major Metals
      2) Byproducts and Coproducts of Minor Metals
   C. Individually Mined Minor Metals
   D. Metals without Geologic Limitations

Table 4 indicates the class or classes into which each minor metal fits. For example, zirconium is in Class B-2, milling byproducts and coproducts of minor metals. This means that it occurs geologically in some mineral species (zircon) associated with one or more other minor metal-bearing mineral species (rutile, ilmenite). All the associated mineral species are mined as joint products, but the zircon is separated from the other minerals during milling, that is, while it is still intact as a mineral but before it is decomposed in order to extract zirconium metal itself.

Table 4 is really a classification of sources of metals rather than a classification of metals as such. It can be used to classify metals only because the currently exploited deposits of most minor metals are restricted to one or two geologic environments.[1] The placing of a metal in a particular slot indicates that the source represented by that slot is the most important one in today's markets and with today's technology. Any metal can move from one slot to another with changes in these conditions. Many metals have been placed in both a main class and (within parentheses) in a subordinate class if there is also substantial production from the source indicated by the latter. A few do not fit well into any class, and here a choice has been made of the most significant association. To some extent the placement of metals is biased in favor of American mining and metallurgical practices. However, since the emphasis in the study is on class characteristics, the placement of a metal can be changed without altering the major theses.

The Classes

Class A-1: Metallurgical Byproducts of Major Metals

At least seventeen minor metals (including most of those consumed by industry before World War II) are extracted from ores of major metals during metallurgical reduction and refining. Such metallurgical byproducts have no separate mining stage of production, and at each plant where they are produced both their value and tonnage are small compared with the production of the major metal with which they are associated. The seventeen minor metals in this class can be subdivided in several ways. For estimation of reserves it is useful to divide these metals according to their affinity for major metals; thus cadmium, germanium, indium, and thallium are all associated with zinc ores; bismuth and antimony with lead ores; and so forth. But because quantity produced comes close to indicating their "importance" relative to major metals in ore deposits, they are subdivided in Table 4 on the basis of annual world production. It is unlikely that net returns from

[1] This is not, however, a chance phenomenon, but is based on the chemical and physical properties of the elements under geologic conditions. The relationship between the classification and these properties can be seen when the classification is superimposed on the Periodic System of the Elements. The metals within any one class tend to be grouped together in the system, hence to possess related properties. David B. Brooks, "The Supply of Minor Metals," *Quarterly of the Colorado School of Mines*, Vol. 58 (January 1963), pp. 11–15.

Table 4. A Classification of Minor Metals

Class A-1: Metallurgical Byproducts of Major Metals

World production greater than 1,000 tons per year:
  Antimony
  Arsenic
  Bismuth
  Cadmium
  Cobalt
World production between 100 and 1,000 tons per year:
  Selenium
  Tellurium
World production less than 100 tons per year:
  Gallium
  Germanium
  Indium
  Platinum-group: platinum, palladium, iridium, osmium, rhodium, ruthenium
  Thallium
  (Scandium from uranium solvent extraction liquors)
  (Thorium from uranium solvent extraction liquors)

Class A-2: Metallurgical Byproducts and Coproducts of Minor Metals[1]

Cesium and rubidium
Columbium and tantalum
Heavy rare-earth group and yttrium
Light rare-earth group and thorium
Hafnium
Rhenium

Class B-1: Milling Byproducts of Major Metals

Radium
Vanadium
(Molybdenum and molybdenum rhenium from porphyry copper deposits)

Class B-2: Milling Byproducts and Coproducts of Minor Metals

Heavy sand deposits:
  Heavy rare-earth group/yttrium
  Light rare-earth group/thorium from monazite
  Columbium/tantalum
  Titanium from rutile
  Zirconium/hafnium
  (Platinum-group metals)
Pegmatite deposits:
  Beryllium from beryl
  Cesium/rubidium from pollucite
  Lithium
  Scandium from thortveitite
  (Columbium /tantalum)
Other deposits:
  (Tungsten from molybdenum ores)
  (Molybdenum from tungsten ores)

Class C: Individually Mined Minor Metals

Mercury
Molybdenum
Tungsten
(Antimony)
(Beryllium from disseminated deposits)
(Cobalt)
(Columbium from pyrochlore)
(Platinum-group metals)
(Heavy rare-earth group from bastnasite)
(Vanadium)

Class D: Metals without Geologic Limitations

Calcium
Sodium

NOTE: Each metal is listed under the class and subclass that represents its most important mode of occurrence in terms of current production techniques. Some metals are also listed within parentheses under a second class and subclass if they are produced from sources represented by that subclass in quantities which are important but secondary relative to the quantities produced from the source indicated by the main listing of the metal. Metals joined by a slash are associated at the indicated stage of production.

[1] The metals noted on each line in Class A-2 are initially produced in natural association either with each other or with another metal in Class B-1 or B-2. The production stage indicated by this class is their separation from one another and their recovery as relatively pure metals.

production of the metals in the two lower tonnage subclasses (under 1,000 tons a year) are large enough to affect major metal production rates to any extent. The high unit value of most of these metals does not make up for their lower rates of production, and in any case is partially offset by higher costs of recovery and purification.

A few of these metals are not exclusively metallurgical byproducts. At various times and places, cobalt and bismuth mines have operated, but they have not been an important source of supply for years. More important, more than half of the world output of antimony is not of byproduct origin. However, byproduct antimony plays a large role in the market in which antimony ores must compete, and, particularly within the United States, is probably the dominant influence. Nevertheless, antimony will also show some of the supply characteristics of individually mined minor metals. Finally, significant quantities of platinum-group metals are produced from two other sources, placer deposits and their own ores. There are no domestic platinum ores, and the proportion of domestic platinum recovery that originates from placers has decreased to less than 30 per cent. But the fact that a fair proportion of the world's platinum is produced from placers and from South African ores means that platinum, like antimony, will have mixed supply characteristics.

Class A-2: Metallurgical Byproducts and Coproducts of Minor Metals

The affinity of most metals in Class A-2 for sister metals is so great that separation has been effected only with modern metallurgical operations. As noted above, such chemically similar metals are produced and consumed in their natural, combined form as much as possible. Moreover, the paired metals in this class are generally found associated in quantities that do not differ from one another by a factor of more than about ten, though in some deposits the difference is much greater. Therefore, most of them can be considered as coproducts, and their supply conditions can be approximated by those of the class (typically Class B-2) in which they were originally mined. However, rhenium and hafnium are always true byproducts and have supply conditions similar to those of the small tonnage subclass of Class A-1. Thus, for the purpose of supply analysis Class A-2 can be neglected.

The case of rhenium deserves special mention, for it is a byproduct of a byproduct. It is produced as a byproduct of the metallurgical processing of molybdenum, but only of that molybdenum which is itself a milling by-product of copper (that is, Class B-1 rather than Class C molybdenum). Strangely enough, individually mined molybdenum does not contain appreciable quantities of rhenium. Inasmuch as rhenium itself does not become a separate metal until the stage of metallurgical reduction and refining, it is placed in Class A-2 rather than Class B-1.

Class B-1: Milling Byproducts of Major Metals

The metals (or groups or pairs of metals) that are joint products at the milling stage are separated from one another while they are still physically and chemically intact within minerals. Class B-1 contains only two metals, vanadium and radium, both of which are recovered from major metals by techniques quite different from those just described for milling byproducts and coproducts of minor metals. Most vanadium is at present a leach product from uranium mills, though in the past it has been mined as a main product. Radium is exclusively a byproduct of leaching uranium ores. Both have supply conditions similar to those for metallurgical byproducts of major metals, and they can therefore be regarded as members of that class for the purposes of supply analysis. Vanadium would be included with the large tonnage byproducts, and radium with the small tonnage. Their similarity to metallurgical byproducts derives both from their association with a major metal and from the fact that the so-called milling operations by which they are extracted are really complex hydro-metallurgical processes.

Class B-2: Milling Byproducts and Coproducts of Minor Metals

Class B-2, the more important of the two milling stage classes, can be divided into three subclasses: heavy sand, pegmatite, and other deposits. The heavy sand (or black sand or placer) subclass refers to extensive areas of beach sand, or less commonly river sand, from which the winnowing action of waves and currents carried away the lighter minerals. A residual concentration of heavier metal-bearing minerals is left, which, if it includes large enough quantities of desired minerals, may become an ore body. Many important heavy sand deposits occur inland where beaches existed in the geologic past.

The pegmatite subclass refers to veinlike masses of minerals that are characterized by exceptionally large crystals. The common white quartz veins are typically pegmatites. Simple pegmatites are common, but valuable metalliferous ones are not. Moreover, metalliferous pegmatites tend to be small and erratic both in vein structure and in metal content.

Finally, lumped into a third class are several geologic associations that permit the recovery of relatively small amounts of certain minor metals from ores of other minor metals during milling.

Most of the metals in Class B-2 are produced from more than a single type of source. Columbium is mined without coproducts from pyrochlore ore, and it is also recovered with tantalum from pegmatite deposits. Production of rare-earth metals is divided between production from sand deposits (monazite) and from an individually mined ore (bastnasite). Similarly, lithium is produced from natural brines as well as from pegmatites.

Class C: Individually Mined Minor Metals

The distinguishing economic characteristic of minor metals mined from deposits in which they are the main products in terms of value is that each must bear the entire cost of its production. Mercury is the best example, for it has been mined extensively as mercury ore for centuries. Price and production of mercury are directly related with no byproducts to share costs or to influence the level of output.[2]

Mercury and tungsten typically occur in small vein deposits. Individually mined molybdenum contrasts with them inasmuch as its ore deposits are large and of relatively low grade. In addition, as already indicated, a large number of minor metals are produced in subordinate amounts from Class C ore sources. In some instances such ore sources are important only during periods of high demand when joint product supplies are inadequate. In other cases, as with pyrochlore sources of columbium today and possibly with disseminated sources of beryllium minerals in the future, ore sources become permanently established, and change the resource outlook for that metal. Ore sources offer the advantage that the quantity and the quality of output can be controlled more readily than with joint product sources.

Class D: Metals Without Geologic Limitations

This utopian-sounding class is an end member in the range of possible supply conditions for all minor metals. The sole, but very restrictive, qualification for entry is a supply so vast that world-wide depletion could not be a problem (in terms of rising cost with existing technology) for at least the next century. At present, sodium and calcium are the only two minor metals that are practically free of supply limitations imposed by the nature of the resource base. Among major metals the extraction of magnesium from sea water is a similar case. And when man begins to extract uranium and other metals from granite and shale, there will be further examples.[3]

Sodium is extracted from the mineral halite, and calcium from the mineral calcite, both of which occur in large monomineralic deposits of rock salt and limestone, respectively, all over the earth. Nevertheless if one were mining for salt or for limestone, there would be a geologic limitation deriving from the size of deposits, their purity, and the like. But the production of sodium or calcium metal consumes such a small proportion of its resource base that it need not be regarded as a resource-depleting process at all.

[2] Edgar H. Bailey and Roscoe M. Smith, *Mercury—Its Occurrence and Economic Trends,* U.S. Geological Survey, Circular 496, Washington, D.C., 1964, pp. 5–7.

[3] It might appear that silicon should fit into this class because of the tremendous deposits of quartz sand. However, the raw material requirements for the production of chemical- or electronic-grade silicon are so stringent that they limit its potential sources very greatly.

4

SUPPLY CONDITIONS
OF THE CLASSES

IMPLICIT IN the classification of minor metals is the idea that each class can be associated with a set of supply conditions distinct from those of other classes. Joint product relationships resulting from the geology underlie much of supply, and, further, supply seems to be the principal determinant of the economic behavior of minor metals. Hence, several supply models, based on joint production, explain the greater part of the price-quantity record of the large number of minor metals. Differences within the classes certainly exist, but for the most part they can be treated as variations on a class theme.[1]

Four classes were found sufficient to cover the major aspects of the supply conditions of minor metals: Metallurgical Byproducts of Major Metals (A-1); Milling Byproducts and Coproducts of Minor Metals (B-2); Individually Mined Minor Metals (C); and Metals without Geologic Limitations (D). The supply conditions of each of these classes will be discussed, but not in turn nor with equal emphasis. Class A-1 represents an end member in two senses: first, the metallurgical byproducts have the most restricted sources of supply; second, their price has the least direct influence over their own production rate. This class is discussed first and is given somewhat more attention than the others because the supply of byproducts has been neglected in mineral economics.[2] Individually mined metals, which have the most direct relationship between price and production, are discussed second, followed by Class B-2 which is intermediate between Class A-1 and Class C in this respect. Class D is considered last because of its tenuous definitional

---

[1] This chapter requires somewhat greater understanding of the techniques of the economist than do other chapters. The initial paragraphs of the discussion for each class summarize the important points.

[2] Although joint production is a nearly universal phenomenon at every stage of production in the mineral industries, it has not received the attention given to joint production in agriculture or manfacturing. The discussion below is based in part on the work of S. V. Ciriacy-Wantrup, "Economics of Joint Costs in Agriculture," *Journal of Farm Economics*, Vol. 23 (November, 1941), pp. 771–818, and T. J. Kreps, "Joint Costs in the Chemical Industry," *Quarterly Journal of Economics*, Vol. 44 (May, 1930), pp. 416–61. Professor Ciriacy-Wantrup distinguished several types of joint production and called the type considered here "process jointness."

relationship with the other classes and because, having wide-open possibilities for supply expansion, it is the other end member in the size of the
available resource.

In the following sections supply conditions are illustrated and analyzed by
means of supply curves. In a sense, each of the four main classes of minor
metals is treated as though it were an industry and the question asked: What
is the typical form of this industry's cost of production curve or, more
broadly, of its supply curve? These curves are industry marginal cost curves,
that is, the sum of the marginal cost curves of all firms producing the metal
in question. Short-run curves for any metal show how much would be offered
for sale by existing producers at alternative prices. Long-run curves show the
price-production relationships after firms have had time to adjust to a new
situation which they do by increasing or decreasing their investment or by
entering or leaving the industry. These long-run supply conditions can also
be expressed by means of shifts in the short-run supply curves.

Demand can be treated by adopting one of the conclusions from Chapter 2,
that it is only possible to define short-run demand conditions for minor
metals. Hence, demand can be visualized in the figures below as steeply
sloping lines that cut through the supply curves at any point (to represent
short-run demand conditions) and that shift their position over time (to
represent longer run conditions). One can analyze the reaction of production
to such exogenous shifts in the position of this inelastic demand curve without
worrying about the cause of the shifts themselves.

With these conditions, and given the severe geologic limitations on expanding the supply of most minor metals, it is not unreasonable that at any time
increasing costs be the penalty for a higher rate of production. This does not
touch on the possibility that costs of minerals go down over time, nor should
the upward sloping curves be taken to imply decreasing quality of ore
deposits over time. The curves that are here shown sloping upward could
shift their position downward (or upward) through time without affecting
the analysis, but this possibility is not within the scope of the study. In other
words, the diminishing returns here in question derive from the law of
variable proportions rather than from the effects of resource depletion.[3]

[3] Diminishing returns may arise from two aspects of a mining firm's operations. First
there are diminishing returns as the annual *rate* of recovery of the desired product is
speeded up in time. This is the aspect that is generally considered in the analyses to follow.
Diminishing returns also arise as the firm attempts to reduce the amount of metal left in
the ground as economically unrecoverable. Similar *level* of recovery decisions must also
be made at the milling and metallurgical stages. See Donald Carlisle, "The Economics
of a Fund Resource with Particular Reference to Mining," *American Economic Review*, Vol.
44 (September, 1954), pp. 595–616.
    A third force of diminishing returns also exists in the mineral industry. This concerns
the long-run industry rate of recovery rather than the short-run firm rate of recovery
just discussed. It results from exhaustion of the richest or most favorably located deposits
first. Whether this force is strongly operative or not depends in part on the reserves-
resource situation.

However, the slope to the industry supply curve does indicate that *at any one time* there will be differences in the qualities of the sources for minor metals, and that the better sources will yield a profit while others are just breaking even.

There is no attempt to be exhaustive in analyzing the supply conditions of minor metals. The response to increased demand is emphasized because it has been the dominant one in the past and because it illustrates the important points. The possibility of a long-run shift from one type of source to another is taken into consideration in the final section of the chapter.

Class A-1: Metallurgical Byproducts of Major Metals

The available supply of byproduct-bearing source materials generally depends more on the rate of production of the major metal with which the byproduct is associated (and thus on the price of the major metal) than it does on the price of the minor metal itself. This relatively inelastic response of production to price changes is the most characteristic feature of byproduct supply. It is an important factor underlying the price record for metals in Class A-1 and also for metals that, as discussed in Chapter 3, have similar supply conditions (rhenium and hafnium in Class A-2 and vanadium and radium in Class B-1).

Byproduct Production

There is a widespread, but ordinarily incorrect, idea that byproducts are free. Actually, they are usually quite expensive to produce. The confusion arises from a semantic difficulty inherent in discussion of most metallurgical byproducts. In some instances the term "byproduct" refers to "waste" materials that result directly from the production of a main product. In other instances the term refers to minerals, metals, chemicals, or other commodities that are recovered from this initially produced byproduct material. The semantic problem can be avoided by using the term "byproduct proper" to refer to initially derived byproducts whenever the distinction between them and more highly processed byproducts is important.

Any byproduct that results from joint production can be classified according to whether separation of a byproduct proper is *necessary* to the production of a marketable main product, and according to whether the separation process is *sufficient* in itself to produce a marketable byproduct commodity. In the mineral industries, the production of most byproducts proper is "necessary." The reasons may be technical, but more commonly they are economic in that the main product is greatly reduced in value if the separation process is not performed. Today's metal specifications require almost complete elimination of most impurities. Sale of off-grade metal and later removal of impurities requires duplication of some costs and hence is not common.[4] Whatever the reason, such "necessary" separation may or may not yield a marketable byproduct. For some byproducts proper, such as the slag produced in smelting, the process of separation is not just necessary, it is also sufficient to produce a commodity that can be sold as ship ballast or road metal with only minor additional costs such as crushing. Others, notably those containing elements that can be profitably recovered, require additional treatment before these elements become marketable as metals or chemicals. Production of the typical byproduct minor metal can therefore be characterized as "necessary *but not* sufficient." Perhaps the example closest to a "necessary *and* sufficient" primary minor metal product is white arsenic; the simplicity of the production process is reflected in its low price of a few cents per pound. From the firm's point of view the output of metal-bearing residues can be considered as "necessary and sufficient" if the firm sells these materials to another firm for recovery of the metals, but such sales are important today only for molybdenum- and cadmium-bearing byproducts.

In cases where impurities have no effect on the value of the main product or actually enhance its value, the potential byproducts are not necessarily separated from the main product. Thus, antimony can either be extracted from lead ores as antimony metal or it can be left in them to yield a primary antimonial lead alloy. Similarly, the tantalum in certain grades of ferro-columbium can either be extracted or left in to form ferrotantalum-columbium. Such "unnecessary" byproducts will be extracted when the net revenue from extraction is greater than that without it. For example, the price at which it becomes profitable to recover antimony metal instead of antimonial lead from impure lead ore is about 40 cents. This example is characterized by production that is both "unnecessary and insufficient." The final possible combination of the two criteria, "unnecessary but sufficient" byproduct production, is represented by very few actual examples.

The criteria of necessity and of sufficiency of byproduct production underlie two quite separate decisions that must be made by management. The first decision is whether to separate a byproduct proper from the main product. The second decision is what to do with any byproducts proper that *are* produced. (The two decisions are not unrelated, for disposition of a byproduct proper will depend in part on the stage of production and the form in which it is produced.) A byproduct proper can be sent to the dump or otherwise disposed of; it can be placed in a stockpile for possible processing at some time in the future; it can be marketed as is; or it can be used as the raw

[4] See S. D. Strauss, "Marketing of Nonferrous Metals and Ores," in *Economics of the Mineral Industries*, ed. Edward H. Robie (New York: American Institute of Mining, Metallurgical and Petroleum Engineers, 1959), p. 284.

material for extracting additional end products. The appropriate decision will depend on many factors including pollution controls and estimates of future demand as well as on the potential net revenue from current production.[5] (This is discussed again in Chapter 7.) The important points here are: first, the cost of a byproduct to the firm cannot be regarded as zero unless it is produced as a "necessary and sufficient" byproduct proper; and second, in all other cases there will be some costs that are directly assessable against the byproduct and that will influence its rate of production.

Long-Run and Short-Run Supply Curves

Joint production implies a relationship between the supply conditions of a main product and of a byproduct, but it does not imply that the short-run and long-run periods of supply will be the same for the two commodities. Although many combinations are conceivable, the supply situations of interest here involve cases in which one or the other of the production characteristics of byproducts forces a wedge between the output of the byproduct and the output of the main product. Thus, alterations in supply conditions of any byproduct whose production is either unnecessary (to produce a marketable main product) or insufficient (to produce a marketable byproduct) can occur independently of those for the main product. Conversely, for strictly "necessary and sufficient" production of byproducts there is no wedge, the output of the byproduct *is* determined uniquely by the output of main product, and there is no distinction between their analytic time periods.

In general the short-run input capacity of the plant for the main product (say, tons of copper concentrates per day) multiplied by the amount of the main product in the input (percentage of copper in the concentrates) sets the maximum short-run output capacity for the main product. The same input capacity multiplied now by the amount of some impurity in the input sets the maximum *long-run* capacity for recovery of that impurity as (or in) a byproduct. In other words, the short-run capacity for the main product also determines short-run capacity for the byproduct proper but only potential capacity for any byproduct extracted from it. Short-run production capacity

[5] In most cases stockpiling will occur in anticipation of higher levels of demand (either because of fluctuations in the demand schedule or because of long-run growth of demand). A compromise solution in effect at most sizable operations is to store byproduct-bearing materials in bulk without additional processing. This procedure minimizes many costs of holding this inventory. Many firms design new metallurgical facilities to allow for possible future extraction of byproducts even though they do not plan to recover them immediately. In recent years stockpiled tellurium-bearing anode slimes proved to be a valuable resource for several firms. *New York Times*, June 16, 1960, p. 45. Similarly, in 1962 when cadmium demand began to exceed the available supply, 20 per cent of production in the United States came from cadmium-bearing flue dust that had been recovered during zinc production in earlier years. *Engineering and Mining Journal*, Vol. 165 (February, 1964), p. 123.

for the byproduct can vary from zero to the full long-run potential depending on the scale of facilities constructed for it; and long-run adjustments in production capacity for the byproduct can be made within these limits without affecting production capacity for the main product.

Refer to Figure 2. Curve *ABC* represents the long-run supply conditions for recovery of a byproduct. The steep lines, labeled *DEF*, D′E′F′, and D″E″F″, represent several of the many possible short-run supply (marginal cost) conditions. The short-run supply curves will intersect the long-run curve. Of course, only one short-run curve is relevant to any particular supply situation. Its position will be determined by the sum of the cost characteristics of all plants in the industry at that time.

The distance *OB′* represents the maximum physical production of the byproduct (100 per cent recovery) as determined by the short-run capacity for the main product. The distance *OA′* represents the per unit cost of production of the minor metal. If it is necessary to extract the minor metal from the major to make the latter marketable, *OA′* will not include this cost of separation but only the cost of further processing of the minor metal. If extraction of the minor metal is not necessary, *OA′* must include both the cost of separation and the costs of further processing of the byproduct (minus any positive or negative change in the value of the main product that results from separation). *AB* is horizontal because, according to the usual analysis of the long run, a plant or series of plants could always be built which would yield the product at minimum cost. However, the *AB* section would not exist for any true "necessary and sufficient" byproduct. Rather, the entire supply situation, short-run and long-run, would be described by a vertical line extending upward and downward from point *B*.

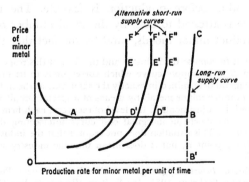

Figure 2. Alternative short-run industry supply curves for metallurgical byproducts of major metals with major metal production capacity constant.

Short-run supply of metallurgical byproducts is relatively inelastic over most of the price range, as indicated by the steep slopes to the short-run supply curves in Figure 2. Production does not respond readily to price changes because of limitations on output variation at individual plants.[6] Nevertheless, some possibilities for varying the output of a byproduct relative to the output of a main product generally do exist. For example, it may pay a firm to obtain a little more of the byproduct by finer grinding of the input (which usually increases byproduct recovery at the expense of main product recovery). Recent metallurgical developments, such as the Imperial smelting process, give promise of increasing the flexibility of plants that treat a variety of ores containing numerous potential byproducts. Thus, the supply of byproducts is not entirely inelastic.

Whatever the opportunities for altering production within the short run, there does come a point beyond which it is impossible to increase production of the minor metal, and after this point (E on Figure 2) the short-run supply curve rises vertically. This is the origin of the extremely inelastic aspect of supply that is so characteristic of byproduct production. Point E is theoretically determined by the minor metal content of the input material when the minor metal plant is operating at maximum capacity. Practically, the limit will be less than this because diminishing returns restrict the level of recovery to something less than 100 per cent.

The long-run supply curve slopes down to Point A indicating that there are economies of scale associated with production of byproduct metals. Economies of scale in the smelting and refining of major metals are very important, at least up to a point.[7] The same situation seems to prevail with byproducts at the metallurgical reduction and refining stage and to be accentuated by the necessity of gathering together large amounts of minor metal-bearing residue before production is feasible. The importance of economies of scale is attested to both by the small number of producers of each of the byproduct minor metals,[8] and by the fact that the major pro-

---

[6] It is important not to confuse the industry and the firm in this respect. For each firm there is also a point B in its supply curve which corresponds to its maximum possible production of minor metal determined in exactly the same way as the industry maximum. In most cases firms produce minor metals by means of a special circuit within the whole plant. Except for possible adjustments in the level of recovery, each firm producing a byproduct generally has its short-run marginal cost curve intersecting the long-run cost curve close to its point B. This situation does not require that the industry supply curve be close to the industry point B, but it does mean that the industry supply curve will probably be inelastic.

[7] Frederick T. Moore, *Industry Organization in Non-Ferrous Metals*, Ph.D. dissertation, University of California, Berkeley, 1951. Joe S. Bain, *Barriers to New Competition* (Cambridge: Harvard University Press, 1956), pp. 53–114, *passim*.

[8] Four companies produce germanium in the United States, three produce bismuth, two produce indium, and only one produces thallium, although the United States is a principal producer of each of these metals.

ducers of metallurgical byproducts put out a variety of them. [9] In other words, it is not profitable (in the long-run sense, including recovery of, and return on, capital) to produce relatively small amounts of metallurgical byproduct metals.

Response of Byproduct Supply to Changed Demand (Major Metal Short Run)

We can now consider the range of possible reactions of minor metal supply to increasing demand short of those associated with any change in major metal production capacity. If demand for the byproduct is initially very low, there may be no production at all (except of "necessary and sufficient" byproducts) because of the diseconomies of small-scale production. Assume that demand is sufficiently high for one or more byproduct recovery plants to be in operation. (At equilibrium the demand curve would pass through the intersection of the short-run supply curve with the long-run supply curve, as at point *D* in Figure 2.) If demand for the minor metal now increases, price will rise sharply (along the same short-run curve) with some resulting increase in production. The increase may come about as management pays more attention to the recovery process. Selective mining may provide concentrates bearing a higher-than-normal content of the minor metal. If the minor metal plant has excess capacity, output may be increased by consuming metalliferous materials from other producers or from a stockpile. [10]

However, in a long-run period for the byproduct, which is the length of time necessary for the construction of new byproduct recovery facilities, there are other possibilities for increasing its supply: (1) expand production with additional or improved recovery facilities at existing plants; (2) expand production with entirely new plants where the metal was not previously recovered at all; (3) expand production of the major metal; and (4) produce the minor metal from new types of sources.

[9] Asarco, the prime example, produces antimony, arsenic, bismuth, cadmium, indium, thallium, selenium, tellurium, and some platinum-group metals, though not all at the same plant to be sure. One report stated that the smelting and refining processes for lead, zinc, and copper concentrates yield more than twenty-one marketable end products, most of them minor metals. Albert J. Phillips, "The World's Most Complex Metallurgy (Copper, Lead, and Zinc)," *Transactions of the Metallurgical Society of AIME*, Vol. 224 (August 1962), pp. 657–73.

[10] Further complications are added if the increased demand for the minor metal causes the price of one of the inputs necessary for minor metal production to rise also. For instance, a firm that bought large quantities of minor metal-bearing materials from other plants and went on to recover the minor metals might find that it had to pay more for these source materials. In these circumstances the industry supply curve is not simply the sum of the firm marginal cost curves, but will be less elastic than the summation curve. However, such external financial diseconomies are neglected here.

Experience indicates that the first two alternatives are the far more likely ones. The additions to capacity will typically provide for substantially increased production and would be represented by a shift of the short-run supply curve to the right, as from DEF to D′E′F′ in Figure 2. Of course, production capacity always expands by discrete jumps, but with minor metals the relative "length" of the jump is greater than with major metals, because of scale economies relative to the size of the industry. The process of supply expansion in this way could theoretically continue until there was a short-run supply curve going right through point B in Figure 2, indicating that the full long-run potential of the byproduct proper was being exploited.[11] If demand still continues to increase, price will increase but not production.

The process of supply expansion by jumps within an inelastic short-run supply situation is shown best by metallurgical byproducts that are extracted from a single main product. Their production record over time will parallel that of the main product (at a much lower volume of production, of course) except that when new or more efficient plants are constructed, it will rise to a higher level and then continue the parallel trend as before. Selenium, for example, is almost exclusively a byproduct of copper refining, although it averages just 0.05 per cent in blister copper. In the early fifties the annual production of selenium increased relative to that of copper, and since then has stayed consistently higher. The figures show that between 1945 and 1951, 0.43 pound of selenium was produced for each short ton of primary refinery copper production. For 1952 through 1961, the same ratio was nearly 0.60. It is unlikely that copper ore suddenly began to contain more selenium. The rise can almost surely be ascribed to the expansion of recovery facilities, since neither a 75 per cent increase in the price of selenium between 1948 and 1951 nor a three-fold increase from late 1953 until 1956 elicited any change in the ratio.

The same sort of effect can be seen when there is a long-run drop in demand. Production is first curtailed by eliminating special operating procedures, such as selective mining. But major cut-backs are achieved by dismantling entire production units; supply is still tied to production of a main product but output is now at a lower level. The change would be shown on a graph by discrete shifts of the short-run supply curve to the left. In

[11] The third force of diminishing returns (cf. footnote 3 *ante*) may determine which firms enter into production. If it is assumed that all firms know the value of byproducts that they could potentially produce, the best sources (presumably lowest cost sources) will be exploited first, and poorer sources brought in later. For byproducts this would result in a shift of the short-run supply curve to the right by progressively smaller and smaller amounts as less and less additional production can be brought in at any price level (given major metal capacity). The long-run supply curve would also slope upward somewhat indicating that resource quality is declining and that owners of poorer source materials have to receive a larger gross return in order to induce them to recover the byproduct.

general, firms with smallest output capacity exit first, which is further
evidence of the importance of scale economies.[12]

Response of Byproduct Supply to Changed Demand (Major Metal Long Run)

Finally, what happens when there is a long-run shift in supply for the
major metal, which has been held so long to a constant level of capacity?
In the usual case the effect of an expansion in major metal production
facilities is to increase the potential supply of the byproduct (i.e., to shift
point B, and the whole long-run supply curve for the minor metal, out to the
right horizontally). A larger minor metal production at the same level of
long-run cost is now possible. But unless facilities for recovering the minor
metal itself are expanded, or unless the facilities were previously operating
below capacity, the minor metal will continue to be produced near the old
rate and under the same short-run supply conditions as before. In contrast,
a decrease in major metal production capacity could force a decrease in
minor metal production. Because the supply of the byproduct proper is
reduced, producers of the minor metal have to operate below capacity.

Class C: Individually Mined Minor Metals

The Mining Firm and Mineral Price Behavior

Metals mined from their own ores, the case in which the metal must bear
the total cost of mining and processing, form the classic problem of mineral
economics. Mining differs from manufacturing in that the production process
is not indefinitely repeatable because a nonreproducible form of capital (the
ore deposit) is consumed. Gray[13] and Hotelling[14] made basic contributions
to the field of mineral economics by describing the modifications that this
factor of exhaustibility makes necessary in the theory of supply and demand
in industry. Herfindahl has summarized their conclusions as follows:

> Under a competitive system, with many mines producing in any one
> year . . . the only situation which will bring equilibrium over time is one

[12] For an example see the reports on arsenic production in *Minerals Yearbook* for the
years following World War II.

[13] L. C. Gray "Rent under the Assumption of Exhaustibility," *Quarterly Journal of
Economics*, Vol. 28 (May 1914), pp. 466–89. Gray showed that a firm producing from an
exhaustible resource can maximize present value by adjusting output not according to the
$MC = MR$ rule but according to the rule that the present value of marginal profits should
be the same in all periods. (This result assumes that ore grade is constant so that cost curves
are the same from one period to the next.)

[14] Harold Hotelling, "The Economics of Exhaustible Resources," *Journal of Political
Economy*, Vol. 39 (April 1931), pp. 137–75. Hotelling generalized Gray's results for the
firm to the whole industry under several states of competition.

in which the value of deposits (i.e., the price of metal in ground) increases at the rate of discount so that the *present value* of the surplus of price over costs of mining and separation is the same no matter in what year the deposit is actually mined.

The quantity produced and consumed will not be the same in each year, of course. It will decline as price rises over time and users shift expenditures to substitutes.[15]

Thus, the theoretical supply conditions of individually mined metals have already been worked out. However, it is still necessary to postulate the particular form of the supply curve that will be applicable.

Short-Run Supply Conditions

Curve *ABC* in Figure 3 is a typical short-run supply curve indicating marginal costs of production for a minor metal mined from ore. The lower, nearly flat section *AB* represents more or less intensive mining of the better deposits of the metal, better in terms of the size of the ore deposit and in terms of all the other physical and chemical characteristics of the ore deposit that can be subsumed under the heading of quality. *AB* includes the possibility of increasing production by using more variable inputs, such as labor, but it does not include the investment of new capital.

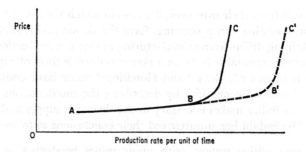

Figure 3. Short-run supply curves for individually mined minor metals.

This lower section *AB* gradually gives way to an upper section *BC* of (probably increasingly) steeper slope. The upper section represents production of metal from a large number of smaller and poorer quality mines, which together produce only small total quantities of metal and which commonly operate only under unusual (wartime) or artificial (price support, tariff) stimuli. The upper section is similar to byproduct supply conditions in its

[15] Orris C. Herfindahl, "The Long-Run Cost of Minerals," *Three Studies in Mineral Economics* (Washington, D.C.: Resources for the Future, Inc., 1961), p. 18.

inelasticity, but for quite different reasons. In the byproduct case, the inelasticity results from the relative lack of value of minor metal production compared with the major metal; in the ore case, it results from higher costs entailed by the physical difficulty of getting additional quantities of metal out of the ground. In other words, the upper part of the ore supply curve represents the effect of diminishing returns from the application of more inputs to a geologic environment that is fixed, at least in terms of our knowledge, in the short run.

This conception of the supply curve for ores requires a twist to the usual definition of the short run. Ordinarily, the short run excludes capital expansion, and indeed curve ABC does exclude expansion of important producers. But it includes the opening (and closing) of the mines represented by BC. These small, high-cost operations are short-lived. Because of their low investment and low employment, they can come into and go out of production readily. In short, they react as if they represented only fuller use of existing capacity rather than new capacity.[16]

If demand for an individually mined metal increases, the rate of production will at first expand quite readily with a mild rise in price. But if the demand is greater than can be satisfied by the better mines in the AB range, price will rise above B, and in the short run the small, high-cost mines represented by the BC range of the supply curve will be able to come into production.

Long-Run Supply Conditions

In the longer run, prices will not remain high for most individually mined metals. High prices induce exploration and development: new deposits are discovered; old ones are expanded; and deposits previously known but undeveloped are brought into production. This long-run expansion of capacity involves costs incurred in the investigation of many deposits, but the process appears to be relatively systematic. That is, exploration is induced by the high profits received by owners of better deposits, and is successful (on the average) in finding additions to supply commensurate with the effort expended.[17] Recent staking rushes for a number of minor metals in antici-

[16] David B. Brooks, "The Supply of Individually Mined Minor Metals and Its Implications for Subsidy Programs," *Land Economics*, Vol. 40 (February 1964), pp. 21–22.

[17] Herfindahl, "The Long-Run Cost of Minerals," *op. cit.*, p. 20. There is disagreement on this point. Some observers feel that unexpected discoveries are more important factors in long-run cost than is systematic exploration, and thus that the cost of finding is not predictable. However, in dealing with the copper industry Herfindahl established a good case for considering the exploration process as a systematic one. *Copper Costs and Prices: 1870–1957, op. cit.*, pp. 40–63. Although it is difficult to draw conclusions for minor metals because the scale of operations is so much smaller, the price and production record, even when changes of source material are involved, seems to indicate that the same is true of them. Certainly whatever randomness exists on the supply side is negligible compared with unexpected developments on the demand side.

pation of increased demand are in themselves partial confirmation of a systematic, if hectic, response in the economist's sense.

The effect of this longer-run effort to increase the supply of individually mined minor metals is generally to increase the production from better quality mines, and often their number. The long-run effect of exploration and development is shown in Figure 3 by a shift of the short-run supply curve from ABC to $AB'C'$. If there are decreasing returns as a result of resource depletion, the slope of the long-run supply curve AB' will be steeper. In any event a new supply-demand equilibrium occurs in the range of prices between A and B', and the quantity demand can again be supplied by the better mines alone. The poorer mines, now represented by $B'C'$, can no longer compete as price drops back to the long-run level; they are forced out of production and do not reappear until a shift in demand forces price above the level of point B'.

The effects described in this model of the production of individually mined minor metals could be established only by matching the costs and production on a mine-by-mine basis. Fortunately, some data of this kind have come out of the various attempts by the federal government to increase production of commodities considered essential to national security. These data, particularly for tungsten concentrate and mercury (both individually mined minor metals) show that the bulk of production came from a few low-cost mines and that a large number of other mines produced only small quantities of ore. Furthermore, most of the smaller producers appeared when the support price was high and then dropped out of production as soon as the support price was reduced.[18] This is just what one would expect if the short-run supply curve was of the ABC form and if the program was viewed by producers as a short-run situation.

Class B-2: Milling Byproducts and Coproducts of Minor Metals

The supply conditions of milling byproduct and coproduct metals are more like those of individually mined metals than those of metallurgical byproducts. It is true that Class B-2 metals lie between the situation in which price must cover the full cost of an increased rate of production and that in which price has little effect on production. However, these metals tend to make up a sufficient part of the value of a deposit to influence the course of production. In addition, because they are recovered at an early stage of production, their output can be adjusted more easily than can that of metals recovered during metallurgical processing.

[18] Brooks, *op. cit.*, pp. 22–24. In this article the cost to the country of maintaining the price high enough to obtain production from high-cost mines was compared with alternative ways of obtaining the same amount of ore.

The short-run supply conditions of both important sub-classes, heavy sand deposits and pegmatite deposits, can be discussed in terms of marginal cost curves similar to *ABC* in Figure 3. But the subclasses differ from one another in the length of the *AB* section of the supply curve (that is, in the extent to which quantity supplied can be increased in the short run) and in the ease of shifting *ABC* to *AB'C'* (that is, in the extent to which supply can be expanded in the long run). In nearly every instance there is a larger number of bigger, better quality, and more easily discovered deposits of valuable sand minerals than of valuable pegmatite minerals. Figure 4 represents two successive short-run supply situations for heavy sand deposits and for pegmatite deposits.

Supply Conditions of Heavy Sand Deposits

Heavy sand deposits occur throughout the world in varying amounts and compositions down to the quality of ordinary beach sand. The larger and richer deposits dominate output because the geology permits economies of scale to be realized. Not only are beach deposits ideal for the use of large-scale equipment but also very large amounts of sand must be processed in order to recover significant quantities of metalliferous minerals. As a result of the importance of scale economies, there are relatively few producers and few producing countries despite the widespread occurrence of heavy sand deposits.

There is every reason to believe that as one heavy sand deposit becomes exhausted, it will be readily possible to find another of equal or only slightly lower quality and size. Hence, the high-cost *BC* sections of the supply curves are shown as very short pieces to indicate that prices in these ranges could not be long sustained. As a matter of fact, the prices of most metalliferous sand minerals are no higher today than they were in 1945 despite the fact that the quantities produced around the world have increased several fold.

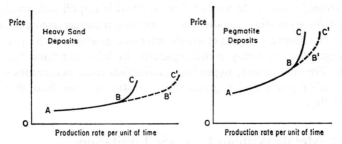

Figure 4. Short-run supply curves for milling coproducts of minor metals. The behavior of the supply curves at low prices in the neighborhood of point *A* varies, but this complication is neglected here.

The curves in Figure 4 do not apply if the value of a metal or its content in the ore is so low that it is really a byproduct rather than a coproduct. The rate of production of the mineral containing that metal then comes to be determined by the value of associated minerals. (Thus, it is likely that the left end of the supply curve turns downward—becomes inelastic—rather than extending in line.) Removal of the byproduct mineral might be a necessary step in production of the main product, but in no case would it be sufficient to produce a primary minor metal product.

Supply Conditions of Pegmatite Deposits

Supply limitations for metals of the pegmatite subclass are more severe than for metals of the heavy sand subclass, and annual world production rates are typically much lower. Not only must the pegmatite vein carry uncommon metalliferous minerals, but if the vein is to yield important quantities of the mineral, it must be large and fairly regular in shape. The areas of the world with pegmatites of this description are few in number. Examples are the multimineral deposits owned by Bikita Minerals in Southern Rhodesia and the lithium deposits owned by Foote Mineral Company in North Carolina, both of which are mined by nonselective quarrying techniques. With increasing demand, the prices of metals of pegmatite origin often go up sharply because the limit of production from the better deposits is quickly reached. Even in the long run, new sources of supply may not be sufficient to bring the price back down to its former level if demand remains at the higher level. The price-production record of most pegmatite minerals contrasts with that of sand minerals; prices have increased substantially since 1945 even though relative production growth has been lower.

Opportunities for scale economies in pegmatite mining are also limited by the fact that the metallic content and properties of many pegmatite minerals are not strongly shown. As a result there is no widely applicable mechanized technique for separating metalliferous pegmatite minerals from one another and from the more common pegmatite minerals, quartz and feldspar. Much of the pegmatite ore today is still separated by laborious hand "cobbing" methods. For this reason, byproduct production from pegmatites appears only at higher prices and is graphically indistinguishable from deposits of lower quality.

Class D: Metals Without Geologic Limitations

Metals without geologic limitations involve the very special situation in which there is an "inexhaustible" source material. The quotation marks indicate that even a source such as the ocean is only relatively inexhaustible.

But it would take centuries of removing metals from the ocean, for instance, for a change in the composition of marine water to be perceptible.

The ideal Class D metal is not only relatively inexhaustible but also relatively ubiquitous and of relatively uniform quality. When all these conditions are satisfied, the value of such a metal "in the ground" (that is, total cost minus the cost of extraction and processing) is zero because there is no need to adjust demand to a supply which is limited by costs increased through depletion or even through depletion of nearby deposits. In other words, these deposits are free goods. There are no diminishing returns arising out of the nature of the resource itself that affect the rate of output. For the few metals in this class, output decisions will be based on plant costs and will not be influenced by natural resource costs. Such extraction processes are distinguished from any other manufacturing process by their name only. Actually, they are comparable to the recovery of bromine from ocean water and of certain industrial gases from the atmosphere. Although essentially a mining process, the recovery of these products is not usually considered to be an extractive industry because of the vast size and availability of the resource base.

The main effect of a potential source of Class D type for some metal is to provide a huge alternative source of supply that sets a firm limit (based on the cost of extraction from this geologically unlimited source) to price increases for metal produced from conventional sources. A synthetic, in effect, creates the same sort of price barrier. In either case, the price of the natural product cannot exceed the cost of production from the alternative source without the latter becoming the major source of supply.

It is this Class D type of source that has been mentioned over and over again as the answer to any problem of metal resource shortages in the future. This sanguine view, perhaps a reaction to the overly gloomy forecasts of the past, is precise but deceptive. There is no question but that a computation of the tonnages of metals contained in a cubic mile of the earth's crust, or of sea water, gives gigantic figures. But it hardly makes sense to consider the bulk of this material as in any practical way available to man or to let this concept obscure the problems of metal supply that may develop in the nearer term future.

The only metals that really fit the requirements for Class D are sodium, calcium, and magnesium. Even the inclusion of sodium and calcium is an approximation. They are treated as if they were extracted from a constantly replenished resource base defined as all the salt and limestone mined in each year. In contrast with magnesium, which is extracted from two sources—sea water and mineral (dolomite)—sodium and calcium come only from the present Class D sources because they are so very much cheaper than any alternative sources.

Intersections of Supply Curves

A supply curve for a minor metal represents its production from only a single type of source. However, as indicated just above, every metal has one or more alternative sources. Each will have its own cost curves, and these curves will intersect. Whenever two supply curves intersect, there is a reversal of the cost relationships between the two sources. Production from the first source does not cease just because it is more expensive at production rates beyond the intersection, but it does become relatively less important than production from the new source. Whether it becomes absolutely unimportant depends upon the slopes of the two cost curves.

Many minor metals are derived simultaneously from two types of sources. Whenever this is the case, marginal costs of production must be approximately the same. Tungsten, for example, is recovered as a milling byproduct at a few mines and, usually in much larger amounts, as an individually mined mineral. At prices below about $18 per short ton unit (20 pounds of tungsten trioxide content) byproduct production is the only source of metal. Ore sources begin to enter production at $18, and they dominate output at almost any price above that level. For example, at $22, byproduct supplies amount to only about one-fourth of annual production.[19]

One particularly common situation involving two types of sources is the transition from a metallurgical byproduct source to a higher cost ore source. Although the supply of a metallurgical byproduct is generally increased by expanding byproduct recovery facilities, ores of these metals may come into production if byproduct sources become strained and prices remain high. When demand conditions relax, or when byproduct supply expands, these higher-cost sources usually drop out of production again. This is exactly the situation that occurs after wars and that has forced domestic antimony and cobalt mines out of production in recent years. In fact, the gap between the cost of byproduct production and the cost of production from the second cheapest source material is generally large. Moreover, in most cases the possibilities of expanding byproduct production without changing the output of major metal are still substantial. For example, during the fifties when extraction of selenium and germanium (both smaller tonnage metallurgical byproducts) from alternative sources seemed a real possibility, it turned out that expanded byproduct supplies equaled demand at prices well below those necessary for production from other sources.

In conclusion, a second type of source that is characterized by higher cost at a low rate of production may limit the opportunity for producing from poor quality deposits of a first type that originally had lower cost. Once the price reaches a certain level the second type of deposit becomes accessible.

[19] Brooks, *op. cit.*, pp. 24–25.

If this additional supply is quite large the upper end of the original supply curve may be of no practical importance. For example, vanadium is largely recovered as a byproduct of uranium mining on the Colorado Plateau, but alternative sources such as titaniferous magnetites contain tremendous quantities of vanadium that could be extracted if price were increased or cost decreased a little.

PART II

THE PRODUCTION
OF MINOR METALS

5

LOCATION AND SCALE
OF PRODUCTION

UP TO this point minor metals have been treated as extractable natural resources and have been classified into groups based on differences in their supply conditions. The emphasis shifts here and for much of the remainder of the study is on the firms producing minor metals and the relationships among them. Consequently the focus also shifts to the metallurgical reduction and refining stage of production.

Although every minor metal goes through three stages—mining, milling, and metallurgical reduction and refining—in one form or another, many of them do not appear as separate commodities until the second or third stage. And it is only at the metallurgical reduction and refining stage that metallic commodities are derived. It is at that stage that metallic elements are separated from their source materials and further processed through the first, commonly marketed product, which may be elemental metal, a metallic compound, or in a few cases an alloy.

Underlying the emphasis on the metallurgical reduction and refining stage of production is the fact that industry is oriented to this stage in the case of almost every one of the thirty-three minor metals included in the study. At this stage the bulk of the metal is commonly in the hands of a relatively small number of firms. These firms collect source material from a larger number of mining operations and produce a relatively uniform primary minor metal product, or, in some cases, several primary products. For certain metals, the primary product is a relatively crude commodity, as with tungsten trioxide or vanadium pentoxide; for most, it is a more highly refined commodity. Whatever the situation with regard to purity, it is generally in terms of this commodity that prices are quoted and statistics are collected. Moreover, the primary product is typically sold in a broad market. At later stages of production there is a substantial increase in the forms in which the metal appears, the processes in which it takes part, and the number of firms handling it. Thus, it is not too much to say that it is the reduction and refining stage that gives minor metals production its character.

51

The orientation of minor metals production to reduction and refining has its origin in a number of factors. Economies of scale exert their greatest influence at this stage. The character of an ore body can severely limit the scale of mine operations, but producers of primary products do not seem to have found a maximum economic size limit for minor metals plants. With byproduct minor metals the collection of large amounts of source material is an important factor underlying economies of scale. With other minor metals continuous processing offers opportunities for winning economies. Even in batch processing, it is generally accepted that the size of the batch is important in reducing average costs.[1] Thus, there will tend to be relatively few primary producers of any given minor metal, a matter to which we will return in evaluating the strength of competition.

It is also at the metallurgical stage that the first substantial inputs of capital (that is, capital that would not be required except for the production of minor metals) are generally needed. Finally, a larger relative increase in value is added at the metallurgical reduction and refining stage than at other stages of production.[2] Whereas ore values are generally on the order of a few cents to a few dollars per pound, the value of primary products typically runs from a dollar to several hundred dollars per pound. Standard forms and shapes of minor metals, in contrast, are usually only two to five times as expensive as the primary products.

In short, the metallurgical stage dominates earlier stages in the recovery process in terms of investment, employment, or value added. And although minor metals require purification, fabrication, or other additional treatment for most applications, these processes are not characterized by the same economies of scale, the joint costs, or the same relative value added as characterize the stage of reduction and refining. Indeed, beyond the stage of producing primary products there is no longer any advantage to considering minor metals as a group. Associations of directly competing materials, such as the ferroalloys or the refractory metals, come to make more sense than the grouping of minor metals.

---

[1] Paul M. Tyler, *General Review of the Titanium Metal Industry*, Materials Advisory Board Report No. MAB–47–SM, 1958, pp. 52–54. S. H. Shelton and staff, *Zirconium—Its Production and Properties*, U.S. Bureau of Mines Bulletin 561, Washington, D.C., 1956, p. 140.

[2] See, for example, the discussion of the beryllium industry by Al Knoerr and Mike Eigo, "Beryllium Update-1961," *Engineering and Mining Journal*, Vol. 162 (September 1961), pp. 87–91. Also, *Report of the Committee on Refractory Metals*, Materials Advisory Board Report No. MAB–154–M (1), pp. 63–64, 125–57, 266.

Geography of the Industry

Location and the Domestic Industry

The location of minor metals production is less dependent on the location of source materials than upon the location of markets. Even at the mining stage there is often considerable latitude in selecting a site because deposits of many minor metals occur rather widely and are not restricted to traditional mining areas. (Lithium, for example, is mined from pegmatites in North Carolina, and zircon from heavy sand deposits in Florida.) Deposits may be small by major metal standards, and yet be adequate as sources for minor metals. If new milling techniques are developed for separating minerals of no obvious metallic character from one another, and then concentrating them, the influence of resource orientation will be further reduced. The use of unconventional sources, such as coal ash, would have the same effect.

Even when the location of a mine is fixed, the location of primary production need not be. Weight loss after mining and milling is commonly not great, and metals plants may wish to draw on several sources. If shipment at some stage of production is necessary, it is typically cheaper to ship concentrated source material to a recovery plant than to interrupt an integrated production sequence that continues through the output of primary products to formed or fabricated articles. The advantage of being close to the mine is further reduced whenever special precautions must be taken in shipping primary products. Cesium metal, for example, must be packed in an inert fluid to avoid reaction with air or water; it is obviously cheaper to ship ore over long distances than to ship metal.

Reinforcing the factors which reduce the dependence of the metallurgical stage on mining are others which increase its dependence on markets. Not the least of these is the fact that the technical knowledge of the producing firms can be more easily transferred to customers if they are near one another. The bulk of the firms consuming minor metals tend to be located in the East (or in California).[3] Adequate power at a reasonable cost and the presence of a skilled labor force are also important location factors. All of these factors help explain the concentration of minor metal plants along Lake Erie. They also explain why molybdenum, mined entirely in the Mountain West, is refined, alloyed, and fabricated in Michigan, New Jersey, Pennsylvania, and Ohio. In short, the output of primary minor metal products, which is now concentrated in the highly industrialized northeast quadrant of the United States, is likely to remain so.

[3] U.S. Bureau of Mines, *Minerals and Metals Commodity Data Summaries*, 1961.

The Foreign Industry and International Trade

Table 5 shows the relationship of domestic consumption to domestic production of both primary minor metal products and minor metal source materials. Table 6 shows the relationship of domestic production and consumption of these commodities to world production. The United States is important as both a producer and consumer of primary minor metal products and as a consumer of minor metals source materials. As a producer of source materials, however, it is significant only for lithium, molybdenum, vanadium, and perhaps a few of the metallurgical byproducts. For a number of minor metals the situation is comparable to that for major metals such as manganese, chrome, and tin, in that domestic mine' production is almost negligible. Actually, the statistics in Table 6 understate our dependence on foreign sources inasmuch as large amounts of minor metals are recovered from imported major metal ores whose minor metal content is not recorded in import statistics. As a result, U.S. producers of primary minor metals rely heavily on imports of source materials. Finally, as Table 6 also shows, the United States is so large a consumer of primary products that it is an importer of them as well.

The influence of foreign production is not limited to actual imports. Because the prices for many minor metals are set in international markets, domestic prices and decisions can be influenced to a considerable degree by foreign production capacity and potential imports. Tables 7 and 8 indicate the influence of these additional factors on producers of minor metals in the United States during the period 1959 through 1962. For few minor metals can such influence be neglected, even when U.S. production is greater than that of the rest of the world put together.

U.S. government policy has been favorably inclined toward imports of minor metal-bearing source materials. Source materials for most minor metals enter this country free of duty (see Table 7). Only tungsten and molybdenum ores bear duties, and the latter is unimportant because we import no "moly." Many underdeveloped countries seem to be well supplied with rich sources for minor metals, and our government has encouraged their development with foreign aid, Commodity Credit Corporation barter agreements, and (formerly) purchase contracts under the Defense Production Act. About the only opposition to these policies has come from those sectors of the strictly domestic mining industry that benefited from the high prices maintained by federal supports in the fifties and that are not associated with integrated facilities that can benefit from the currently lower prices. In most instances the primary producers have successfully defended free trade in minor metal source materials.[4]

[4] *Oil, Paint, and Drug Reporter*, October 12, 1959, p. 4 (on cobalt ores). *American Metal Market*, May 17, 1962, p. 17 (on cobalt ores); May 2, 1962, p. 14 (on beryl); and February 10, 1964 (divergent views on tungsten ores).

Table 5.   Relationships Between Domestic Production and Consumption of Minor
Metal Source Materials and Primary Products, 1960

Metal	Mine production as percentage of:			Primary products production as percentage o f primary products consumption
	Consumption of mined products	Primary products		
		Production	Consumption	
Antimony	—	15	11	75
Arsenic[1]	—	40*	8*	19
Beryllium	4	n.ap.	n.ap.	—
Bismuth	—	50*	40	92
Cadmium	—	40*	40*	100
Calcium	n.ap.	n.ap.	n.ap.	—
Cesium	0	0	0	100
Cobalt	76	100*	18	18*
Columbium	0	0	0	>100
Gallium	—	100*	100*	>100*
Germanium	—	52	52	100
Hafnium	n.ap.	n.ap.	n.ap.	100
Indium	—	50–60*	10*	20–25*
Lithium	45*	n.ap.	n.ap.	>100
Mercury	n.ap.	n.ap.	n.ap.	65
Molybdenum	152	157	214	136
Platinum group[2]	n.ap.	40*	3*	7
Radium	0	0	0	0
Rare-earth metals[2]	—	40*	40*	100*
Rhenium	—	>100	>100	>100
Rubidium	0	0	0	100
Scandium	—	>100	>100	≥100
Selenium	—	—	—	75
Sodium	n.ap.	n.ap.	n.ap.	>100
Tantalum	0	0	0	75
Tellurium	—	—	—	111
Thallium	—	—	—	≥100
Thorium	—	—	—	100
Titanium	40	n.ap.	n.ap.	97
Tungsten	58	n.ap.	78	110
Vanadium	92	147	360	246
Yttrium	n.ap.	n.ap.	n.ap.	≥100
Zirconium	52	n.ap.	n.ap.	≥100

NOTE: In Appendix C the procedures used for collection and presentation of minor metal statistics are discussed in more detail. Percentages neglect recovery ratios from metal in ore to consumption of metal products so that the table overstates the extent of self-sufficiency. The metal products used to make calculations are those listed in text Table 3. All sources of mined production that could be indentified were used in calculation.
  * Estimate based on information other than that in Appendix tables.
  n.ap.   not applicable.
  — Data insufficient to make an estimate.
  [1] 1959 data.
  [2] Totals.

Table 6.   Relationships Between Domestic Supply and World Production of Minor Metals
in Source Materials and in Primary Products, 1960

Metal	As per cent of world mine production		As per cent of world primary products production	
	U.S. mine production	U.S. consumption of mined products	U.S. primary products	
			Production	Consumption
Antimony	2	—	18[1]	24[1]
Arsenic†[2]	—	—	11	57
Beryllium	4[3]	104[3]	—	—
Bismuth†	—	—	27	29
Cadmium	—	—	41	41
Calcium	n.ap.	n.ap.	10–15*	25
Cesium plus rubidium	0	—	75*	—
Cobalt†	5[4]	6[4]	5*	27
Columbium†	0	88	60–80*	—
Gallium	—	—	50*	>100
Germanium†	—	—	45	45
Hafnium	n.ap.	n.ap.	90*	—
Indium	—	—	20*	—
Lithium†	30	66	70*	65*
Mercury	n.ap.	n.ap.	14	21
Molybdenum	76	50	50	35
Platinum group total	2*	n.ap.	4	61
Radium	0	—	0	—
Rare-earth metals	————classified————		65*	—
Rhenium	—	—	50*	50*
Rubidium	——————————included with cesium——————————			
Scandium	—	—	80*	—
Selenium†	—	—	32	43
Sodium	n.ap.	n.ap.	85	—
Tantalum	——incl. with Cb——		70–90*	—
Tellurium†	—	—	70	62
Thallium	—	—	50–75*	—
Thorium	——————————————classified——————————			
Titanium†	8	21	58	60
Tungsten	10	18	—	—
Vanadium†	79[5]	87[5]	—	30*
Yttrium	—————included with rare-earth metals—————			
Zirconium†	25	47	80*	—

NOTE: In Appendix C the procedures used for collection and presentation of minor metal statistics are discussed in more detail. The primary products used to make calculations are those listed in text Table 3. All sources of mine production that could be identified were used in calculation. Unless statistics were available (or could be recalculated) on a comparable basis, no attempt was made to arrive at a percentage.

† World production figure excludes U.S.S.R. and/or other non-Western nations.
* Estimate based on information other than that in Appendix tables.
n.ap. not applicable.   — Data insufficient to make an estimate.
[1] For calculation of this ratio, world mine production figure was adjusted for 92 per cent recovery.
[2] 1959 data.
[3] Beryllium content of world beryl production estimated for comparison with domestic production.
[4] For comparison with world production, 78 per cent recovery assumed.
[5] Metal content of world vanadium ores estimated for purpose of comparison with domestic production.

Table 7. Influence of Foreign Production on U.S. Producers of Sources for Minor Metals[1]

Minor metal	U.S. producers dominate world market	Foreign producers have: mild influence on U.S. market	Foreign producers have: strong influence on U.S. market	Foreign producers dominate world market	1960 U.S. duties (¢/lb.)[2]
Antimony			X		0
Arsenic			X		0
Beryllium				X	0
Bismuth			X		0
Cadmium			X		0
Calcium		Not applicable			(3)
Cesium				X	0
Cobalt				X	0
Columbium				X	0
Gallium		X			0
Germanium			X		0
Hafnium[4]				X	0
Indium		X			0
Lithium			X		0
Mercury		No significant trade			0
Molybdenum	X				$0.30
Pt-gr. metals				X	0
Radium				X	0
RE metals		X			0
Rhenium		No significant trade			0
Rubidium				X	0
Scandium			X		0
Selenium			X		0
Sodium		Not applicable			(3)
Tantalum				X	0
Tellurium			X		0
Thallium		X			0
Thorium			X		0
Titanium			X		0
Tungsten			X		$0.50
Vanadium	X				0
Yttrium		Included with rare-earth metals			0
Zirconium			X		0

[1] U.S. producers (or alternatively foreign producers) were considered to dominate the world market if they could set prices with little or no regard for the quantities of output that might come from outside the U.S. (alternatively, from within the U.S.). Foreign producers were considered to have a strong influence if imports made up any significant proportion of domestic supplies at any time between 1959 and 1962, or if domestic prices changed more or less in accord with prices in London and other centers of metal trade. Foreign influence was considered mild if U.S. exports made up a significant proportion of foreign consumption, or if international price differentials tended to persist for more than a few weeks.

[2] For metallurgical byproduct metals the duty refers either to ores or to the intermediate metallurgical products from which they are extracted. Such material may also be subject to a duty on any base metal content.

[3] Duties of 3¢ per hundred pounds of hydrated lime and 17¢ per hundred pounds of bulk salt have not been listed; there is little trade, and material that is imported is not used as a source of metal.

[4] High-hafnium zircon.

Table 8. Influence of Foreign Production on U.S. Producers of Primary Minor Metal Products[1]

Minor metal	U.S. producers dominate world market	Foreign producers have: mild influence on U.S. market	Foreign producers have: strong influence on U.S. market	Foreign producers dominate world market	1960 U.S. duties (¢/lb. or ad valorem)[2]
Antimony			X		M:2¢ / Oxide:1¢
Arsenic			X		White:0
Beryllium		X			M:21% / A:22½% / C:12½%
Bismuth			X		M:1⅞% / A:22½%
Cadmium			X		M:3¾%
Calcium		X			M:17½%
Cesium		No significant trade			M:21% / C:10½%
Cobalt				X	M:0 / Oxide:4¢
Columbium		X			12½%
Gallium		X			12½%
Germanium			X		11%
Hafnium		No significant trade			12½%
Indium			X		M:10½%
Lithium		X			M:25% / C:12½%
Mercury			X		M:25¢
Molybdenum	X				25¢ + 7½%
Pt-gr. metals			X		0
Radium				X	0
Rare-earth metals		X			Misch:$1 / Mixed salts:30%
Rhenium		X			10½%
Rubidium		No significant trade			10½%
Scandium		No significant trade			10½%
Selenium			X		0
Sodium	X				M:21%
Tantalum	X				12½%
Tellurium		X			M:10½% / C:12½%
Thallium		X			10½%
Thorium		No significant trade			M & A:12½% / C:35%
Titanium		X			M:20% / A:12½ to 25%
Tungsten		X			42¢ + 25%
Vanadium		X			A:12½% / C:40%
Yttrium		No significant trade			10½%
Zirconium	X				12½%

[1] U.S. producers (or alternatively foreign producers) were considered to dominate the world market if they could set prices with little or no regard for the quantities of output that might come from outside the U.S. (alternatively, from within the U.S.). Foreign producers were considered to have a strong influence if imports made up any significant proportion of domestic supplies at any time between 1959 and 1962, or if domestic prices changed more or less in accord with prices in London and other centers of metal trade. Foreign influence was considered mild if U.S. exports made up a significant proportion of foreign consumption, or if international price differentials tended to persist for more than a few weeks.

[2] M = metal; C = compounds; A = alloy. Duties on alloys refer only to minor metal content; such products may also be subject to a duty on any base metal content. All tariffs refer to "unmanufactured" material. Some duties have been reduced since 1960. Tariff when unqualified refers to all forms of the metal.

American dominance in the output of many primary minor metal products can be expected to be short-lived because of the growing awareness in other industrialized countries of the commercial possibilities in this field. (Few plants to produce primary minor metals have been built in underdeveloped countries, probably owing to the capital-intensive nature of the production techniques, the lack of assured markets for the products, and the fact that such markets as do exist are confined almost entirely to industrialized nations.) Belgium has always been a major producer of large tonnage metallurgical byproducts and is increasingly becoming so for smaller tonnage ones. Japan is a strong second to the United States in the output of sponge titanium (if the Soviet Union is not second) and is exporting a number of other minor metals. Canada is a large producer of selenium, tellurium, cobalt, platinum, and lithium. The Soviet Union is also paying increased attention to minor metals.[5] Again, some of these developments, such as Japanese titanium, can be traced directly or indirectly to inducements associated with our national security program. Whatever their origin, further competition from these countries is to be expected both in product markets and for source materials. By the same token, domestic industry has been quick to take advantage of commercial opportunities and technologic advances originating in other countries. International trade in minor metals, however, is not without restrictions. The United States, along with many other countries, has erected substantial tariff walls around primary metals. Present tariff rates for primary minor metal products average 10 to 12½ per cent (Table 8). Inasmuch as the duties on imported source materials (aside from tungsten concentrates) are zero, the primary metals sector gets the full benefit of these tariffs, and there is little question but that they have been effective in restricting international trade in minor metals. Although reciprocal agreements have succeeded in reducing the level of tariffs on minor metals by about 50 per cent since the 1930's, there is now strong opposition to reducing them any further.[6]

Size and Growth of Output

The production of minor metals does not form an industry in the ordinary sense. Nevertheless, it is useful to have some idea of how large—perhaps it would be better to say "how small"—a production effort is devoted to the mining and processing of minor metals. Annual value of production for minor metals as a group is estimated in the following section. Because value

[5] *Metal Bulletin*, September 4, 1964, p. 12; *Mining Journal*, September 18, 1964, pp. 2–3.

[6] *American Metal Market*, May 2, 1962, p. 14, and February 27, 1964, p. 1. See also the statement of Senator Alan Bible of Nevada favoring retention of the present tariff on titanium metal. *Congressional Record*, October 15, 1964, pp. A5319–21.

data for minor metals are scarce, the quantitative results are fairly rough, but they do indicate that major metals quite dominate the production of either ores or primary products. Another quantitative impression of the scale of minor metals production can be obtained from the year-to-year changes in the tonnage of minor metals production. The weighted index numbers, presented in the second section below, show that, although minor metals as a group do not play a large role in the metals industry, their production has grown much more rapidly since 1945 than has that of major metals.

Value of Minor Metal Production

Despite the attention paid to minor metals since 1945, it is clear that in both physical and value terms minor metals production is still very small, compared with major metal production. Indeed, only about 2.75 per cent of the $417 billion of 1960 national income in the United States can be considered as originating in *all* domestic metal mining, milling, smelting, and refining.[7] Roughly half of this $12 billion would be accounted for by iron and steel, and most of the remainder by the five major nonferrous metals. Minor metals cannot account for more than a small fraction of this figure. They hold an equally subordinate place in Europe where the five major nonferrous metals are reported to represent over 95 per cent of nonferrous metals consumption.[8]

A better idea of the place of minor metals in the minerals sector of the economy can be obtained by separating the value of production data for source materials from that for primary products. In 1960 the value of all metallic minerals produced in the United States in the form of ore or rock (that is, before the recovery of primary products) was $2 billion. In comparison, the value of minor metal ores produced in 1960 was estimated at $133 million, nearly 65 per cent of which is accounted for by molybdenum concentrates.[9] The figure of $133 million does not include metallurgical byproducts of major metals because it accounts only for the value of production of those minor metals for which mine production data are recorded by the Bureau of Mines. However, it includes the total value of production for those minerals used both as a source of a minor metal and as a mineral

[7] The source for the figures in this section, unless otherwise indicated, is the chapter "Review of the Mineral Industries" in the 1960 *Minerals Yearbook*, Tables 1 and 25.

[8] Statement by W. Casper of Metallgesellschaft to Metals Division of the European Economic Community Commission in Florence, May, 1962, as quoted in *Metal Bulletin*, October 5, 1962, p. 17.

[9] Value of production data for minor metals are rarely given by the Bureau of Mines because of disclosure problems. This figure was determined in the process of calculating the index numbers in Table 9. It equals 1960 production for the selected source materials listed in Appendix Table 2a weighted by 1960 prices before the figures were recalculated to a base of 100.

product. The value of minor metals in all source materials, including by-product sources, could not have exceeded $150 million.[10]

An order of magnitude estimate for the value of production for all metals at the metallurgical stage can be obtained from the monthly average sales of primary metals in 1960.[11] This figure, multiplied by twelve, gives $25.8 billion in total, or $9.7 billion exclusive of iron and steel. A roughly comparable figure for the value of primary minor metal products produced in 1960 is about $230 million, neglecting only a few minor metals whose value of production could not be estimated with comparable accuracy. Even if all were included, the value could not have exceeded $325 million.[12]

Certainly if minor metals do not compare favorably with major metals in terms of value, they are going to compare even less favorably in terms of tonnage, employment, or other measures of scale. It is obvious that the course of minor metals production will have a negligible effect on even the mining and metal sectors of the economy. In contrast, the course of major metal production will have a deep and long-lasting effect on the production of minor metals.

An Index of Physical Volume of Production

Most people have simply assumed that minor metals are a "growth industry," or at least as much of a "growth industry" as is likely to develop in metals, without attempting to detail this growth. There are two measures of metals production but neither indicates much about minor metals alone.

[10] The value of metallurgical byproducts in ores has probably been overestimated for two reasons. First, aside from gold and silver, most small tonnage byproducts have no value in the ore. Except in special instances, such as high germanium content, metallurgical plants do not pay any premium for the minor metal content of purchased concentrates. Second, in a few cases (e.g., antimony and arsenic in copper ores) the minor metal content carries a negative price; metallurgical plants exact a penalty from the seller if his ores contain more than a stipulated low percentage of these metals.

[11] *Survey of Current Business*, Office of Business Economics, U.S. Department of Commerce. See Table of "General Business Indicators." The monthly averages as given were $2.15 billion in total and $1.34 billion for iron and steel.

[12] The total value of the selected primary minor metal products shown in Appendix Table 2b is $132 million exclusive of molybdenum and $195 million including molybdenum. In addition, although data are not available to make estimates over time, estimates for the value of production in the single year 1960 could be calculated for beryllium, calcium, cesium plus rubidium, columbium plus tantalum, gallium, hafnium, indium, platinum metals (other than platinum and palladium), rhenium, and thallium. In each case the output of metal products listed in Table 3 was multiplied by an average or quoted price for that product in 1960. The result of including all these metals gave the figure of $230 million. Rough estimates for the value of production of tungsten, sodium, and a few other metals raised the estimate to about $300 million, from which figure the $325 million maximum was estimated. The difference between $325 million and $150 million cannot be used as a measure of value added because of the importance of trade in source materials (especially exports of molybdenum concentrates) and the use of some source materials as mineral products.

The first index is the Federal Reserve Board Index of Industrial Production, which is given for a few years since 1945 in Part A of Table 9. The index clearly shows the relative stagnation both in the "Metal Mining" category, which measures the production of metals and minerals at the mine, and in the "Primary Metals" category, which measures the production of metal products from both ore and scrap at metallurgical plants. Neither is closely related to minor metals because each is overwhelmed by the major metals.

The second index is a little more helpful, although it includes only mine production and not primary metals production. This is the Bureau of Mines Index of the Physical Volume of Mineral Production, which is shown for the same years in Part B of Table 9. The "All Minerals" index shows about the same growth as does the FRB index. However, several of the subcategories are of more interest than are the totals. In 1960 the "Other Nonferrous Metals" category (which includes bauxite and magnesium as well as the minor metals, cadmium, mercury, and platinum) stood at 168 on a 1947–49 base, thus showing far greater growth than the group as a whole. This rate of growth was exceeded only by "Construction Nonmetals." Moreover, lithium, classed here as a metal, is included in the category of "Chemical Nonmetals," which shows the third highest growth rate.

Neither index gives minor metals sufficient weight for one to form an adequate impression of the production changes that have occurred in recent years. However, statistics collected on the production over time of selected minor metals formed a basis on which an acceptable index could be constructed. Appendix Tables 2a and 2b give, respectively, the statistics for mine production and for primary products production in the years 1945, 1950, 1955, and 1960. By aggregating figures for the values of production at 1960 prices, index numbers for the two series were determined and are shown in Part C of Table 9.[13]

Two cautions are necessary. First, the index numbers were computed by summing across the metals at one stage of production in one year, not by summing across metals at different stages of production. Second, the base years for the minor metal indexes are different from those for the FRB index and the Bureau of Mines index, both of which use average unit values during 1947–49 as weights. Because a number of minor metals were not even produced during the forties, and for other reasons discussed in Appendix D, 1960 prices were chosen as weights for the minor metal production data. Then, in order to make all the indexes readily comparable, the arithmetic average of the weighted minor metals production figure for 1945 and that for 1950 was set equal to 100 in order to approximate to the 1947–49 base used for the federal indexes.

[13] The technique used is described in more detail in Appendix D.

Table 9.   Indexes of Mining and Metal Production in the United States, 1945 to 1960

Item	1945	1950	1955	1960
A. Federal Reserve Board Indexes of Industrial Production[1] (1947–49 average = 100)				
Total industrial production	107	112	139	164
Minerals	92	105	122	128
Metal mining	n.a.	108	110	134
Manufacturing	110	113	140	163
Primary metals[2]	n.a.	115	140	115
B. U.S. Bureau of Mines Indexes of the Physical Volume of Mineral Production[3] (1947–49 average = 100)				
All minerals	92.0	102.6	119.0	122.6
Metals, total	95.2	108.8	115.0	104.6
Nonferrous base metals	97.4	109.0	106.8	105.1
Other nonferrous metals[4]	142.3	113.9	194.0	167.6
Nonmetals, total	70.2	116.1	161.0	192.2
Chemical nonmetals	82.2	112.9	146.2	162.7
C. Indexes of Production of Minor Metal Source Materials and of Minor Metal Products[5] (Average of 1945 and 1950 = 100)				
Mine production:				
Excluding molybdenum	95	105	304	253
Including molybdenum	101	99	241	238
Metal products production:				
Excluding molybdenum	116	84	167	196
Including molybdenum	114	86	153	177

[1] Indexes for 1955 and 1960 can be found in recent editions of the "Federal Reserve Bulletin." The indexes for 1950 recalculated on the 1947–49 base can be found in the December, 1953, issue of the "Bulletin," pp. 1298–1323.

[2] The term primary metals as here used refers to production of metal products from both virgin and scrap materials.

[3] "Minerals Yearbook," Vol. 1, 1956 and 1960.

[4] Includes bauxite, magnesium, cadmium, mercury, and platinum.

[5] The method of calculating these indexes is described in Appendix D.

The products included in the index for mine production account for nearly all sources that were produced as ores of minor metals, but none that were metallurgical byproducts of other metals. The index probably represents over 75 per cent of the value of minor metals in all source materials and close to 100 per cent of the value of minor metal ores. The index in Part C of Table 9 indicates growth of physical output through 1960 of about 2½ times over the 1945–1950 average. Because nearly two-thirds of the value of mine products is accounted for by molybdenum, the index is shown with and without molybdenum. The inclusion of molybdenum has a damping effect. Nevertheless, there clearly was a peak in the output of mined products between 1950 and 1960. These changes can be attributed to the Korean War and the various federal programs of stockpiling and price supports, which were at their height in the middle fifties.

The index for the output of primary products is also shown on Part C of Table 9. The metals included in this index represent 60 to 80 per cent of the total value of output of minor metals in 1960. (Molybdenum trioxide accounts for about one-third of the included values.) The index fell from 1945 to 1950 but then it subsequently recovered to about twice the 1945–50 average. This post-1950 recovery results from growth of output in specific groups of metals. (See Appendix Table 2b.) Metallurgical byproducts of major metals have fared no better than major metals. In most cases their output has declined absolutely since 1945 or since the conclusion of a federal support program. The only significant exceptions are germanium, selenium, and tellurium, all metals that have found important new uses. Vanadium, which is similar to a metallurgical byproduct, has also increased in output because it is associated with uranium rather than with copper, lead, or zinc. Most of the growth of the index in recent years has come from metals of the milling byproducts and coproducts class or from individually mined minor metals. For all the fluctuations in the rates of consumption, the increase in use or the initiation of use of such metals as titanium, lithium, and the rare-earth metals has been impressive.

In sum, while the advent or the curtailment of certain federal programs has caused peaks in the production record of particular minor metals, the index numbers substantiate the impression that, over all, the output of minor metals has grown significantly, but not spectacularly, in recent years. Moreover, they indicate a steadier growth rate for the production of primary minor metals than for the production of their source materials. In part this is a result of the fact that federal programs involving source materials turned out to have no lasting effect on production. In part it is also a result of the construction of the index since the sources of metallurgical byproducts of major metals, which would have exerted a steadying influence on the index, could not be included for lack of data. And finally, it is a result of the general and growing dependence of the United States upon foreign sources of raw materials and of our specialization in the highly technical phases of industry that are capable of producing and consuming large quantities of the newer minor metals. For the same reasons, the output of minor metals will probably continue to grow during the sixties, and the bulk of the growth will probably be in the production of primary metals and more highly processed products rather than in the output of source materials.

6

THE FIRMS PRODUCING MINOR METALS

ALMOST EVERY major industrial concern in the United States works with some minor metal at some stage in its operations. However, since the key to an understanding of the structure of minor metals production lies in the metallurgical reduction and refining stage, we restrict our attention in this chapter to the less than one hundred firms that produce primary products. To a large extent, we shall view them descriptively and leave evaluation of the structure for the following two chapters. Readers interested in general outlines rather than details may prefer to skip directly to the summary at the end of this chapter.

Industry, Firm, and Minor Metals

Table 10 lists nearly every firm found to be producing minor metals at the metallurgical reduction and refining stage of production, plus the parent firms of joint ventures. It includes all the firms that produced important primary minor metal products (as listed in Table 3) over the period 1958 to 1962. Only the smaller mercury and tungsten producers, which were too numerous to list, have been deliberately omitted.

Some of the 83 firms listed in Table 10 produce only small amounts of minor metal in any year and consume most of what they do produce in their own industrial processes. For example, besides the five firms recovering vanadium from uranium ore and ferrophosphorous slag, three firms recover small amounts of byproduct vanadium from chromite ore and consume it in their own plants. Similarly, a number of firms recover small amounts of antimony compounds from ore for use in their production of matches, fireworks, and other products. It was found that the main characteristics of minor metals production could be brought out more clearly if such firms were put to one side. In Table 11, therefore, the unimportant producers are omitted, as are the parent firms of joint ventures. The remaining 60 firms are not necessarily either large in size or large producers, but each produces

Table 10. U.S. Producers of Primary Minor Metal Products, 1958-62

Firm (1)	Important producer of minor metals in class:[1] (2)	Small producer of: (3)	Recent exit from production of: (4)	Also mentioned in connection with (5)
Allegheny Ludlum Steel	See TMCA	—	—	—
Allied Chemical	—	Se	—	Ti
Aluminum Co. of America	A-1	—	—	—
American Metal Climax	A-1, B-1, C	—	—	Li, W
American Potash & Chemical	A-2, B-2	—	—	—
American Scandium	—	RE, Sc, Y	—	—
American Smelting & Refining	A-1	—	—	Mo
American Zinc, Lead & Smelting	A-1	—	—	—
The Anaconda Co.	A-1	—	—	Cb, Mo, Be, Bi, V
F. W. Berk	—	V	—	—
The Beryllium Corp.	B-2	—	—	—
J. Bishop & Co. Platinum Works[2]	C	—	—	—
Brush Beryllium	B-2	—	—	—
Bunker Hill	A-1	—	—	Co
Carborundum Co.	A-2, B-3	—	—	—
DeCoursey-Brewis Minerals	C	—	—	—
Dow Chemical	A-2, B-2	—	Ti	Th, Na, Mg
E. I. du Pont de Nemours & Co.	A-2, B-2, D	—	Si	Zr
Duval Sulphur & Potash[3]	B-1	—	—	—
Eagle-Picher	A-1	—	Si	—
Ethyl Corp.	D	—	—	Ca
Fairmount Chemical	A-2, B-2	—	Sc	—
Fansteel Metallurgical	A-2, B-2	—	—	W
Foote Mineral	B-2	Sb	Si	Hf
W. R. Grace & Co.	A-3, B-3, C	—	Si	—
Harshaw Chemical	—	Sb, Cd	—	—
Hercules Powder	—	V	—	—
Howe Sound	A-1, C	—	Co	—
Hummel Chemical	—	Sb	—	—
Kawecki Chemical	A-2, B-2	Se	—	Ge, Te
Kennametal	A-2	—	—	W
Kennecott Copper	A-1, A-2	—	—	Mo, Ti, Cb, Te
Kerr-McGee Oil Industries	B-1	—	—	—
Kleber Laboratories	—	RE	—	—
Lithium Corp. of America	B-2	—	—	Be
Lunex Co.	—	RE, Y	—	—
Mallinckrodt Chemical Works	B-2	—	Cb-Ta	Si
McGean Chemical	—	Sb	—	—
Merck & Co.	C	—	—	—
Metal and Thermit[4]	—	Sb	—	—
Metallurgical Resources	—	Co	—	—
Michigan Chemical[5]	A-2, B-2	—	—	—
Mineral Concentrates & Chemical	B-2	—	—	—

Footnotes at end of table.

Table 10 (continued)

Firm (1)	Important producer of minor metals in class:[1] (2)	Small producer of: (3)	Recent exit from pro- duction of: (4)	Also mentioned in connection with: (5)
Minerals Engineering	B-1, C	—	—	—
Minerals Refining	—	FeCb	—	—
Molybdenum Corp. of America	B-1, B-2, C	—	—	FeW, W, Cb
National Distillers & Chemical	A-2, B-2, D	—	Cs	—
National Lead	A-1. See also TMCA	—	Co	FeZr, Ti, Zr
National Research	A-2, B-2	—	Zr	—
National Zinc[6]	A-1	—	—	—
New Idria Mining & Chemical	C	—	—	—
The New Jersey Zinc Co.	A-1	—	—	Ti
Nuclear Corp. of America	A-2, B-2	—	—	Sc
Pfizer (Chas.) & Co.	D	—	—	Mg
Phelps Dodge	A-1	—	—	Mo
Philadelphia and Reading	—	FeCb	—	FeW
Pittsburgh Plate Glass	A-2, B-2	V	—	Ti
The Pyrites Co.[7]	A-1	—	—	—
Rare Metals Corp. of America	C	—	—	—
Republic Steel	—	Mo	—	—
Richardson-Merrell	—	Mo	—	—
Rocky Mountain Research	—	—	Cs, Rb	—
St. Eloi	—	—	RE, Sc	—
St. Joseph Lead	A-1	—	—	—
Sharon Steel	See National Distillers	—	Ti, Zr	—
S. W. Shattuck Chemical	—	Mo	V	—
Sherwin Williams	—	Cd	—	—
Shieldalloy	—	FeCb, FeTa-Cb	—	FeV
Sonoma Quicksilver Mines	C	—	—	—
Stauffer Chemical	A-2, B-2	—	—	Ti
Stepan Chemical	A-2, B-2	—	RE	—
Sun Oil	C	—	—	—
Sunshine Mining	A-1	—	—	—
Susquehanna	B-1. See also Miner- als Engineering.	—	—	—
Sylvania Electric Products[8]	A-1	—	Si	W
Titanium Metals Corp. of America[9]	B-2	—	—	—
Transition Metals & Chemicals	—	FeCb	—	—
Union Carbide	A-2, B-1, B-2, C, D	—	Ti	FeZr, Sc, FeW, FeV, Si
United Refining & Smelting	A-1	—	—	—
U.S. Smelting, Refining & Mining	A-1	—	—	—
Vanadium Corp. of America	B-1, B-2	—	—	FeZr, FeV
Vitro Corp. of America	A-2, B-2	—	—	—
Wah Chang	A-2, B-2, C	—	—	—
Yuba Consolidated Industries	C	—	—	—

Footnotes at end of table.

Notes for Table 10.

NOTE: Firms listed in the table are those which produce primary minor metal products, i.e. the products shown in column 2 of Table 3. Mines, mills, and firms producing only the products listed in column 3 of Table 3 are not listed. With two exceptions noted below the list is thought to be complete as of the period 1958 to 1962 (including firms that ceased or initiated production in those years). However, a few small producers (particularly those which consume their own output) may have been inadvertently omitted. The two general exceptions are mercury and tungsten. Because of the large number of small mercury metal producers, only those firms producing over 1,000 flasks in 1960 have been listed. The tungsten producers listed include those firms with a chemical processing plant or with an exceptionally large mill. Subsidiaries are in most cases listed under the parent corporation if they are consolidated on statements. Affiliates are listed separately.

[1] Firms producing metals in Class A-1 are listed again on Table 11, Part A. Firms producing metals in Classes A-2, or B-2 are listed again on Table 11, Part B. Firms producing metals in Classes B-1, C, or D are listed again on Table 11, Part C.

[2] Affiliated with Johnson, Matthey & Co., Ltd., a London firm (not listed).

[3] United Gas Corporation owns 75% of stock; unconsolidated.

[4] In December, 1962, Metal and Thermit was merged into American Can Co. and became M&T Chemicals Division.

[5] Velsicol Chemicals owns 35% of stock.

[6] Controlled by International Minerals-Metals.

[7] Subsidiary of Rio Tinto-Consolidated Zinc Corp., Ltd., a London firm (not listed).

[8] Nonconsolidated subsidiary of General Telephone and Electronics Corp. (not listed).

[9] Jointly owned by National Lead (50%) and Allegheny Ludlum (50%).

a relatively significant amount of at least one primary minor metal product—say, 5 per cent of U.S. production. Part A of Table 11 includes the firms producing metallurgical byproducts of major metals; Part B includes the firms producing milling byproducts and coproducts of minor metals (and the metals recovered from them at the metallurgical stage); and Part C includes the firms producing metals of the other three classes. A firm will appear more than once if it produces minor metals in classes that are listed on separate parts of Table 11.

Which Industries Produce Minor Metals?

The firms producing minor metals are listed below according to the general product line that accounted for the largest proportion of their revenue in 1960.[1]

[1] For the most part, firms were classified on the basis of information reported in *Standard and Poor's Standard Corporation Records*. For example, W. R. Grace and Company reported the following breakdown of its consolidated sales and revenues: chemicals, 41 per cent; steamship operations, 12 per cent; foreign operations, 21 per cent; petroleum refining and petrochemicals, 14 per cent; and all other, 12 per cent. The largest share was chemicals (even neglecting chemical phases of other categories), and accordingly Grace was listed as a chemical concern. If a firm did not reveal sales information, the description of the firm given by *Standard and Poor's* or by Bureau of Mines sources was used as a guide.

| | *Number of Firms Listed in:* | |
	Table 10	*Table 11*
Mining and milling	3	3
Primary nonferrous metals	16	15
Metal fabricating and alloying	10	5
Petroleum and gas	2	2
Iron and steel	3	0
Chemical	25	14
Electrical	1	1
Minor metals	10	7
Other manufacturing and miscellaneous	13	13
Total	83	60

The "minor metals" group is made up of firms that produced minor metals as the largest part of their activities, and to some degree it cuts across the lines established by the other industries. Only ten firms in all, most of them small producers, could be found that fell into this group. Few firms specialize in minor metals; rather, most firms produce them as a relatively small part of their over-all operations.

Three industrial groups dominate production of minor metals regardless of whether all producers (Table 10) or just the more important producers (Table 11) are included: primary nonferrous metals, chemical, and other manufacturing and miscellaneous. The large number of firms in the other manufacturing and miscellaneous group is deceptive. In the case of almost every firm in this group, the division of the firm which is actually producing minor metals could be separately classified as either a chemical firm or a nonferrous metals firm. National Lead (Texas Mining and Smelting Division) and Pittsburgh Plate Glass (Columbia-Southern Chemical Division) are examples. Other firms in the manufacturing and miscellaneous group have changed character over time or have expanded into new fields. For example, the Eagle-Picher Company is one of the oldest domestic mining firms, but it now secures 75 per cent of its gross revenue from manufacturing and only 25 per cent from mining and smelting.

Few producers of minor metals come from the petroleum industry, the iron and steel industry, or the other industries, particularly if attention is restricted to the more important producers. However, the small number of firms in the mining and milling industry is explained by the fact that most such firms were eliminated by definition. Only an unusual firm that sold the bulk of its output in the form of ore or concentrates but that also produced some primary minor metal product would be so listed.

Table 11. Major U.S. Producers of Primary Minor Metal Products by Metal Class

A. Metallurgical Byproducts of Major Metals (Class A-1)

Company	Plant location	Iron	Gold	Copper				Lead		Zinc			Al
		Co	Pt-group	Se	Te	As_2O_3	Sb[1]	Bi[2]	Cd	Ge	Tl	In	Ga
Alcoa	Bauxite, Arkansas												✓
Amax	Carteret, N.J. (Cu-Zn smelter and refinery)		✓	✓	✓								
	Blackwell, Okla. (Zn smelter)									✓			✓
Asarco	Denver, Colo. (Globe cadmium plant)								✓				
	Corpus Christi, Texas (Zn smelter)										✓		
	Perth Amboy, N.J. (Pb and Cu refinery)		✓						✓				
	Omaha, Nebraska (Pb smelter and refinery)						✓	✓					
	Tacoma, Washington (Cu smelter and refinery)				✓	✓	✓	✓				✓	
	Baltimore, Maryland (Cu refinery)			✓	✓								
American Zinc	East St. Louis, Illinois (Zn smelter)								✓	✓			
Anaconda	Perth Amboy, N.J. (Cu refinery)		✓	✓	✓								
	Great Falls, Montana (Zn plant)			✓					✓			✓	✓
	Anaconda, Montana (Cu smelter and refinery)				✓	✓							
Bunker Hill	Kellogg, Idaho (Zn plant)								✓				

Firm	Location										
Eagle-Picher	Henryetta, Okla. (Zn smelter)										
	Miami, Okla.	(√)									√
Howe Sound	Garfield, Utah (cobalt refinery)								√		
Kennecott Copper	Garfield, Utah (Cu smelter)		√					√			
National Lead	Laredo, Texas (antimony smelter)						√				
	Fredericktown, Mo. (nickel-cobalt plant)	(√)									
National Zinc	Bartlesville, Okla. (Zn smelter)								√		
New Jersey Zinc	Palmerton, Pa. (Zn smelter and refinery)								√		
	Depue, Ill. (Zn smelter and refinery)		√						√		
Phelps Dodge	Laurel Hill, N.Y. (Cu refinery)	√				√					
The Pyrites Co.	Wilmington, Delaware	√									
St. Joseph Lead	Herculaneum, Missouri (Pb smelter)								√		
	Josephtown, Pa. (Zn plant)						√		√		
Sunshine Mining	Kellogg, Idaho (antimony refinery)						√			√	
Sylvania Electric	Towanda, Penna.					√					
United R & S	Chicago, Illinois			√		√					
USSR&M	E. Chicago, Indiana (Pb refinery)		√		(√)	√					
	Midvale, Utah (Pb smelter)										

Footnotes at end of table.

√ Firm produces metal.   (√) Firm recently ceased production of metal.

B. Milling Byproducts and Coproducts of Minor Metals (Class B-2) and Associated Metals (Class A-2)

Company	Plant Location	A-2										B-2								
		Re³	Hf	Cs-Rb	Cs	Rb	Cb	Ta	In-div. RE	Y	Th	Heavy RE and Y	Light RE and Th	Cb-Ta⁴	Ti	Zr	Be	Cs-Rb	Li	Sc⁵
American Potash	Searles Lake, Calif., and Henderson, Nevada									✓	✓	✓								
	W. Chicago, Ill.								✓	✓	✓		✓							
	San Antonio, Texas				✓													(✓)	✓	
Berylco	Reading, Pennsylvania					✓											✓			
Brush	Elmore, Ohio																✓			
Carborundum Co.	Parkersburg, W.Va.		✓													✓				
Dow Chemical	Midland, Michigan			✓														✓		
du Pont	Newport, Delaware					✓									✓					
Fairmount Chemical	Newark, N.J.			✓										✓				✓		(✓)
Fansteel	N. Chicago, Ill., and Muskogee, Okla.						✓	✓						✓						
Foote Mineral	Sunbright, Va., and Exton, Pa.										✓		✓							
Grace	Pompton Plains, N.J.																		✓	
	Boyertown, Penna.						✓	✓												
Kawecki Chemical	Fort Washington, Penna.						✓	✓										✓		

Kennametal	Latrobe, Penna.	..	..	..	..	..	√	√	..	..	..	..	..	..	..	..	..	..	..	..
Kennecott Copper	Waterbury, Conn.	√	..	..	..	..	..	..	..	..	..	..	..	..	..	..	..	..	..	..
Lithium Corp.	Bessemer City, N.C.	..	..	..	..	..	..	..	..	..	..	..	..	..	..	..	..	..	√	..
Mallinckrodt Chemical	North St. Louis, Missouri	..	..	..	..	..	..	..	..	..	..	(√)	(√)	(√)	..	..	..	..	..	..
Michigan Chemical	St. Louis, Michigan	..	..	..	..	..	..	..	√	√	..	√	..	..	..	..	..	..	..	√
Mincon	Loveland, Colorado	..	..	..	..	..	..	..	..	..	..	..	..	..	..	..	√	..	..	..
Moly Corp.	Washington, Penna.	..	..	..	..	..	..	..	..	..	..	..	..	√	..	..	..	..	..	..
National Distillers	Cincinnati, Ohio	..	..	..	(√)	..	√	..	..	..	..	..	..	√	..	..	..	..	..	..
	Ashtabula, Ohio	..	√	..	..	..	..	..	..	..	..	..	..	..	√	√	..	..	..	..
National Research	Newton, Mass.	..	..	..	..	..	..	√	..	..	..	..	√	..	..	(√)	..	..	..	..
Nuclear Corp.	Phoenix, Ariz.	..	..	..	..	..	..	..	√	√	..	√	√	..	..	..	..	..	..	..
PPG	Pensacola, Fla.	..	√	..	..	..	..	..	..	..	..	..	..	..	..	√	..	..	..	..
Stauffer Chemical	Niagara Falls, N.Y.	..	..	..	..	..	√	√	..	..	..	..	..	√	..	..	..	..	..	..
Stepan Chemical	Maywood, N.J.	..	..	..	√	..	..	..	..	..	(√)	..	(√)	..	..	..	..	√	√	..
TMCA	Henderson, Nevada	..	..	..	..	..	..	..	..	..	..	..	..	..	√	..	..	..	..	..
Union Carbide	Niagara Falls, N.Y.	..	..	..	..	..	√	√	..	..	..	..	..	√	(√)	..	..	..	..	..
VCA	Cambridge, Ohio	..	..	..	..	..	..	..	..	..	..	..	..	√	..	..	..	..	..	..
Vitro Corp.	Chattanooga, Tenn.	..	..	..	..	..	..	..	√	√	√	√	√	..	..	..	..	..	..	√
Wah Chang	Albany, Oregon	..	√	..	..	..	√	√	..	..	..	..	..	..	..	√	..	..	..	..

√ Firm produces metal. (√) Firm recently ceased production of metal.

Footnotes at end of table.

C. Individidaly Mined Metals (Class C) and Other Classes (B-1, and D)

Company	Plant Location	B-1		C					D	
		V	Mo	Hg	Mo	W	RE	Pt-Group	Ca	Na
Amax	Grand Junction, Colo.	✓	…	…	…	…	…	…	…	…
	Climax, Colo.	…	…	…	✓	…	…	…	…	…
J. Bishop and Co.	Malvern, Penna.	…	…	…	…	…	…	✓	…	…
DeCoursey-Brewis	Aniak District, Alaska	…	…	✓	…	…	…	…	…	…
du Pont	Niagara Falls, N.Y., and Memphis, Tenn.	…	…	…	…	…	…	…	…	✓
Duval Sulphur & Potash	Pima, Arizona	…	✓	…	…	…	…	…	…	…
Ethyl Corp.	Baton Rouge, La., and Houston, Texas	…	…	…	…	…	…	…	…	✓
Howe Sound	Vance County, N.C.	…	…	…	…	✓	…	…	…	…
Kerr-McGee Oil	Shiprock, New Mexico	✓	…	…	…	…	…	…	…	…
Minerals Engineering	Garfield, Utah	✓	…	…	…	…	…	…	…	…
	Salt Lake City, Utah	…	…	…	…	✓	…	…	…	…
Moly Corp.	Washington and York, Penna.	…	✓	…	…	…	✓	…	…	…

Firm	Location
National Distillers	Ashtabula, Ohio
New Idria	Fresno, Calif.
	Idria, Calif.
Pfizer	Canaan, Conn.
Rare Metals Corp.	Washington County, Idaho
Sonoma Quicksilver	Sonoma County, Calif.
Sun Oil	Humboldt County, Nev.
Susquehanna	Garfield, Utah
Union Carbide	Rifle and Uravan, Colo.
	Niagara Falls, N.Y.
	Bishop, Calif.
VCA	Durango, Colo.
Wah Chang	Glen Cove, N.Y.
Yuba Consolidated	Yuba River, Calif.

√ Firm produces metal.   (√) Firm recently ceased production of metal.

[1] Antimonial lead is produced by alloying lead and antimony at other plants.

[2] Bismuth is also partially refined by Anaconda (IS & R Division).

[3] Also produced at the University of Tennessee, Knoxville, Tennessee.

[4] Includes producers of mixed columbium-tantalum products and producers of ferrocolumbium and ferrotantalum-columbium.

[5] Includes scandium recovered from uranium as well as from thortveitite.

Firms from the primary nonferrous metals industry account for 25 per cent of the concerns listed in Table 11. Two types of behavior seem to be common among these firms. One type is exemplified by Asarco, Inco, and many European firms, which have expanded their operations by recovering more and more of the metallurgical byproducts contained in nonferrous ores.[2] They have extended the intensive margin of exploitation by utilizing their initial input to the fullest. The other type of behavior is exemplified by Amax, Anaconda, and Kennecott, which have actively explored for, and produced from, ores of entirely different character. Amax, for example, is a complex company that holds interests in lithium, tungsten, and cesium-rubidium as a result of investments in African and Canadian mining companies. Also, as the world's largest producer of molybdenum, it has gone into the whole field of refractory metals.[3] Similarly, Anaconda, which first departed from its traditional nonferrous interests by investing in uranium mines, has since invested in beryllium, columbium, and other minor metals.[4]

There are almost twice as many firms from the chemical industry producing minor metals as from any other industrial grouping. It is possible to distinguish two types of chemical firms that produce primary minor metal products. The first type is the very large chemical firm, such as du Pont or Union Carbide, which has investments in the output of a number of minor metals.[5] The second type is the specialty chemical firm, which produces less familiar chemicals or salts of rare elements. Among such firms here are

[2] See Albert J. Phillips, "The World's Most Complex Metallurgy," *Transactions of the Metallurgical Society of AIME*, Vol. 224 (August, 1962), pp. 657–73; "Inco Now Recovers 14 Elements in Canadian Operation," *INCO Magazine*, Vol. 26 (July, 1956), p. 18; Jim Morrison, "How Vielle-Montagne Integrates Byproducts Recovery in Pb-Zn Plants," *Engineering and Mining Journal*, Vol. 162 (December 1961), pp. 92–98.

[3] See the report of the address by Frank Coolbaugh, President of Amax, to the Chicago Society of Security Analysts in the *American Metal Market*, April 27, 1962, p. 1, which was significantly entitled "Amax's New Image: To De-Emphasize Some Base Metals."

[4] Anaconda is a major investor in St. Lawrence Columbium and Metals Corp. of Canada, the world's largest producer of columbium minerals, and has announced plans for a joint exploration program in Quebec. *New York Times*, October 29, 1961; *American Metal Market*, January 19, 1963, p. 1. It also owns beryllium deposits in Nevada and has erected a pilot ore processing plant. *Chemical Week*, April 11, 1964, p. 27. In 1963 it purchased the controlling interest in General Astrometals, a beryllium fabricating and alloying concern. *American Metal Market*, January 7, 1963, p. 1. An official of Anaconda is on the Board of Fansteel Metallurgical, a leading tantalum producing and fabricating firm.

[5] Major chemical firms have also placed themselves in a position to reap benefits from substitution against metals as well as substitution toward metals. They are large producers of metal-like plastics and metal-plastic composites. See *American Metal Market*, June 12, 1962, p. 8; and *Barron's*, "Plastics vs. Metals," Vol. 41 (September 4, 1961), pp. 11–12. Metal-like ceramics and cermets are generally produced by a different group of firms, largely in the electrical industry and the abrasives industry. The 1963 merger of National Research and Norton Company created the first major firm specializing in both minor metals and ceramics.

Mallinckrodt Chemical, Kawecki Chemical, and Michigan Chemical. These firms have experience in specialized techniques, such as solvent extraction and ion exchange, and they are capable of using these processes on a variety of difficult-to-extract metals. Kawecki and Stauffer, along with some of the very large firms, have also diversified into metallurgical techniques, such as electron beam melting and high-temperature fabrication of refractory metals.

The blurring of the distinction between the chemical industry and the nonferrous metals industry began before World War II. Since the war, the entry of chemical firms into the production of metals, and to a smaller extent the entry of metals firms into the production of chemicals and petrochemicals, has been an important factor in the growth of these firms. This tendency to produce across product lines was quite naturally carried over into minor metals, some of which respond better to chemical extraction and some of which respond better to pyrometallurgical or electrometallurgical extraction.

Which Industries Produce Which Minor Metals?

It is not enough to know which industries produce minor metals. Only when we know which industries produce *which* minor metals does the structure of production begin to take real form. The three parts of Table 11, and Table 12 which is compiled from them, give a good part of the answer. Above all, it clearly brings out the distinction between those minor metals produced principally by firms in the nonferrous metals industry and those produced principally by firms in the chemical industry.

Primary nonferrous metals firms dominate the production of metallurgical byproducts of major metals (Class A-1). Of the nineteen producers listed in Part A of Table 11, fifteen are in the primary nonferrous metals industry (see Table 12). Three others are among those firms in the miscellaneous grouping that retain mining and smelting interests. The remaining producer, Sylvania, is in the electrical industry and produces germanium from imported concentrates. In addition, each of the few nonferrous metals firms that does produce minor metals in another class, also produces metallurgical byproducts of major metals.

Firms from the chemical industry dominate the production of milling byproducts and coproducts of minor metals (Class B-2) and the metallurgical byproducts of minor metals that are derived from them (Class A-2). Of the twenty-nine producers listed in Part B of Table 11, eleven are chemical firms, and six are firms in the miscellaneous grouping that have large chemical divisions (Table 12). Thus, 60 per cent of the firms producing metals in these two classes can be considered as chemical firms. The one nonferrous metals firm is Kennecott, which produces rhenium, the only

Table 12.   Firms Producing Classes of Primary Minor Metal Products by Industrial Grouping

Industrial grouping	Number of firms in the grouping	Class of primary product[1]					
		A-1	A-2	B-1	B-2	C	D
Mining and milling	3	0	0	1	0	3	0
Primary nonferrous metals	15	15	1	1	0	1	0
Metal fabricating and alloying	5	0	2	2	3	2	0
Petroleum and natural gas	2	0	0	1	0	1	0
Iron and steel	0	0	0	0	0	0	0
Chemical	14	0	10	2	11	1	4
Electrical	1	1	0	0	0	0	0
Other manufacturing and miscellaneous	13	3	7	1	8	3	1
Minor metals	7	0	0	0	5	2	0
Total number of firms producing in class	60	19	20	8	27	13	5

[1] A-1: Metallurgical byproducts of major metals,
A-2: Metallurgical byproducts and coproducts of minor metals,
B-1: Milling byproducts of major metals,
B-2: Milling byproducts and coproducts of minor metals,
C: Individually mined minor metals,
D: Metals without geologic limitations.
SOURCE: Table 11.

metal in these two classes that is recovered from a nonferrous metal.[6] Four other firms, predominantly interested in the sister metals, tantalum and columbium, are from the metal fabricating and alloying industry, and two more are metallurgically-oriented firms from the miscellaneous group, both of which produce zirconium and hafnium. The five remaining producers of milling byproducts and coproducts of minor metals are among those firms that derive the greatest part of their revenue from producing and marketing minor metals. By isolating this class, it can be seen that firms specialized to minor metals produce a very limited range of products. Of the ten such firms that could be found (see Table 10), eight produce exclusively one or more metals in this class: three produce rare-earth and yttrium metal products,[7] three produce beryllium metal products, one produces lithium, and one titanium. (Titanium should not really be included; the one titanium firm, TMCA, although listed separately, is not an independent firm but an affiliate of National Lead and Allegheny Ludlum Steel.) The two other minor metals firms produce mercury, an individually mined metal.

[6] As described in Chapter 3, rhenium is a byproduct of that molybdenite that is itself a byproduct of copper.

[7] None of these three firms was an important enough producer to be listed on Table 11. Two of them have gone out of business since 1962.

A much more heterogeneous group of firms is found producing the metal products listed in Part C of Table 11. The eight firms producing milling byproducts of major metals (Class B-1) represent six different industries. Thirteen firms from seven industries produce individually mined minor metals (Class C). Three firms are from the mining industry and three from the miscellaneous group. Also, the two minor metals firms that did not appear above, both mercury mining firms, are included here. Finally, production of metals that do not have geologic limitations on the rate of production (Class D) is exclusively the domain of chemical firms or chemical divisions. (This is also true of the magnesium that is produced from sea water and of most products extracted from sea water, the atmosphere, or other essentially unlimited sources.)

Behind the heterogeneity of this group lies the fact that the firms produce metals that are not associated at the natural resource level. Also, many of the metals listed are recovered from individually mined sources that are unique. For example, Moly Corp. is recovering rare-earth products from the only bastnasite deposit known to be under exploitation in the world. In contrast, the minor metals listed in Part A of Table 11 are all associated with a few major metal ores, and those listed in Part B commonly occur together in the same deposit or in the same mineral. To use the same example, other rare-earth mineral processors use monazite, generally found in heavy sand deposits, as a source material and are listed together with processors of rutile, zircon, and other sand minerals in Part B.

The influence of a common source material carries further. It also goes far toward explaining why many firms produce more than one minor metal. Thus, nonferrous metals firms typically recover several minor metals from the same ore. Likewise, because sister metals commonly occur in pegmatite or heavy sand minerals, most firms that produce milling coproduct metals also go on to produce the metallurgical byproducts and coproducts that are derived from them. However, these relationships will not remain static if source materials change. For example, so long as the main source of columbium was columbite, which also contains appreciable quantities of tantalum, most firms that recovered columbium also recovered tantalum. But now that pyrochlore, an individually mined source of columbium containing little tantalum, is available, more firms may elect to produce only the former. (The reverse is not true; no source of tantalum comparable to pyrochlore has been discovered so most tantalum firms still produce columbium.)

Diversification among minor metals is, therefore, rather common. A minimum measure of the extent of diversification can be obtained by noting that 24 of the 60 firms producing primary minor metals produce metals in more than a single class (Table 11). Two-thirds of these produce metals in two different classes, five produce metals in three classes, and one (Union

Carbide) produces metals in five different classes. Diversification among minor metals is of course more extensive than this because many firms produce more than one metal, but all within one class. On the other hand, as will be discussed further below, firms tend to limit diversification to metals that have a common source or are produced by similar techniques.

There are no representatives of the minor metals group among the firms producing across class lines. Upon looking more closely one finds that the few firms specialized to the production of primary minor metals are highly specialized indeed, most of them producing just a single minor metal. However, in only two cases, mercury and beryllium, is the production of a minor metal dominated by such specialized firms.

The fact that mercury is produced by specialized companies is the result of its unique properties, which permit nearly pure mercury to be produced by a simple process right at the mine site. Although the long-standing dominance and reported high profits of a few large Spanish and Italian mines attest to the fact that there are economies of scale in mercury recovery, most American deposits are too small for such economies to be of much importance.[8] As a result, the larger mining and metallurgical concerns in this country have not engaged in mercury production. Moreover, consuming firms have also stayed out of mercury production because their needs are generally periodic rather than continuous, and because these needs could usually be supplied by brokers or dealers. Consequently, the field has been left to small firms, which by almost any other definition would be mining firms, but which here fall willy-nilly into the primary products sector.

The situation is quite different with beryllium. Three firms, all specialized to this one metal, produce primary products, but two of them, the Beryllium Corporation and Brush Beryllium, dominate the industry at this and later stages of production. (The third firm, Mincon, processes lower grade Colorado beryllium ores into commercial-grade compounds. These chemicals are sold largely to the glass and ceramics industry as a substitute for direct consumption of beryl when closer control of content is required; they do not compete directly with the products of Brush and Berylco.) Most beryl, the principal source material for beryllium, is a byproduct at pegmatite mines, but domestic production is small and the mining firms are not integrated.

One must turn to the history of the beryllium industry for suggestions as to why it is the only primary minor metal produced entirely by specialized firms. First, production of beryllium-copper alloy (a nonsparking, hard alloy, which until recently was the dominant product of the industry) is quite complex and began in the early thirties when few concerns were interested

---

[8] U.S. Tariff Commission, *Mercury (Quicksilver)*, Report on Investigation No. 32 Under Section 332 of the Tariff Act of 1930, Washington, D.C., 1956, pp. 7–8, 26–27, 28, 62–72.

in new ventures. Second, large amounts of the source material, beryl, were not produced by any going concern in the normal course of its operations. As will be emphasized in the next chapter, had a beryllium source material been producible as a byproduct by one of the integrated metals or chemical firms, the stimulus for that firm to enter the beryllium field would have been strong. Third, the technology for producing beryllium in various metallic forms was originally developed in Germany, and Berylco began production on the basis of German patents. These patents were tightly controlled, and a Congressional Hearing in 1939 developed extensive evidence on the monopoly character of the beryllium industry.[9] Brush Beryllium was able to break into the field on the basis of a new and cheaper process in the 1930's, and it strengthened its position by developing products other than beryllium-copper. By the time that beryllium came to the attention of the AEC and the aerospace industry, the two established companies had the capacity to produce large amounts of alloy and the technical know-how to begin producing ductile beryllium metal. They had to expand, and they got ample government assistance for expansion, but it was clearly cheaper to graft this expansion onto Brush and Berylco than for the government to encourage the formation of new companies.[10] Although two firms have recently entered the beryllium industry, only one of these new entrants, Anaconda, is a truly diversified and integrated concern that could bring major financial resources into a competitive struggle.[11]

Vertical Integration

The prevalence of diversification of producing firms at the metallurgical reduction and refining stage of production was noted above. What about the vertical integration of the same firms throughout all stages of production for a single minor metal? Table 13 shows that most firms producing primary minor metals are integrated into either mining or further processing, but not into both.

[9] Temporary National Economic Committee, *Development of the Beryllium Industry*, Hearings, 76th Congress, 1st Session, pt. 5, 1939, pp. 2011–2163 and 2276–304.

[10] It is argued with strong justification that this method of inducing expansion generated competition in the beryllium industry by forcing Brush and Berylco to compete in each other's traditional markets. U.S. Attorney General, *The Beryllium Industry*, Report Pursuant to Section 708 (e) of the Defense Production Act of 1950, as Amended, 1960, pp. 43–44.

[11] Anaconda holds a controlling interest in General Astrometals, a beryllium fabricating firm, and has also become interested in utilizing the low-grade disseminated beryllium ores in Utah and Nevada. The other entrant in the beryllium industry is Beryllium Metals and Chemicals, a subsidiary of Lithium Corporation. This firm is smaller than the existing beryllium producers but like them was specialized to the production of one minor metal. In an attempt to diversify to other metals, Lithium Corporation has started refining beryllium scrap using a newly developed Japanese process. *American Metal Market*, May 21, 1963, p. 13; *Chemical Week*, March 23, 1963, pp. 22–23.

Table 13. Vertical Integration in the Production of Primary Minor Metal Products

Form of integration[1]	Firms producing primary minor metal products in Class:				
	A-1	B-1	A-2 and B-2	C	D
Total number of firms producing in the indicated class	19	8	29	13	5
Number of firms integrated into mining sources for the indicated primary products	15	8	6	11	n.ap.
Number of firms integrated into further processing or consumption of the indicated primary products	5	4	27	7	5
Number of firms integrated in both ways listed immediately above	4	4	6	6	1

[1] Integration implies use of the integrated facilities, not just ownership. Firms were considered as being integrated if they utilized the mining or processing facilities over the period 1958 to 1962.

SOURCE: Table 11, Bureau of Mines information, and corporation annual reports.

Firms producing metallurgical byproducts of major metals as well as those producing milling byproducts of major metals generally produce their own source materials. Being integrated into mining for the production of major metals, these firms are also integrated into mining for the byproduct minor metals that they decide to recover. In contrast, except for the production of high-purity metals or ferroalloys, they do not generally process minor metals beyond the primary product stage.

Firms producing milling byproducts and coproducts of minor metals exhibit the reverse behavior. Integration into further processing is nearly universal, but integration into mining is unusual. In fact, only six firms have mining operations; four of them produce lithium minerals, and two produce rare-earth minerals. Since production of milling byproducts and coproducts is not integrated into mining, neither is the production of the metallurgical byproducts and coproducts of minor metals (Class A-2), which are derived from them.

Firms producing individually mined metals (Class C) are divided; somewhat more of them are integrated into mining operations than into further processing of the primary products. Finally, all of the firms producing metal from a geologically unlimited source (Class D) consume the primary product they put out. (Source materials for most Class D metals are not mined in the usual sense of the word. Is a firm that recovers magnesium from sea water integrated into mining?)

Vertical integration turns out to be a very important factor in analyzing decisions on entry into the production of minor metals, and is discussed again in the next chapter.

Size of Firms

To the extent that information was available, firms producing primary minor metals were grouped according to their total assets at the end of 1960 into ten size groups which form an approximately geometric progression (see Table 14). Since there was no attempt to have the groups contain equal numbers of firms, the even distribution that resulted indicates a far greater proportion of large firms engaged in minor metals production than is typical for other fields (see the first column of Table 14). Minor metals production cannot be said to be the domain of large firms, but it is obviously a field in which large firms have taken a strong position.

Table 14. Size and Age of Firms Producing Primary Minor Metal Products by Class of Product

Assets and age	All U.S. corporations[1] (thousands)	Number of firms in each size or age class				
		Table 10	Table 11	Part A, Table 11	Part B, Table 11	Part C, Table 11
Assets (thousands):[1]						
Less than $2,500	1,108	4	4	0	2	2
$2,500–$4,999	14	1	1	0	0	1
$5,000–$9,999	8	3	3	1	2	0
$10,000–$24,999	6	9	9	0	7	2
$25,000–$49,999	2	10	8	2	5	3
$50,000–$99,999	1	9	7	3	2	4
$100,000–$249,999 }	0.8	{ 7	4	2	2	0
$250,000–$499,999 }		{ 7	6	5	0	2
$500,000–$999,999 }	0.6	{ 6	5	1	4	2
$1,000,000 and over }		{ 6	5	2	3	2
Unknown	—	21	8	3	2	3
Total	1,140	83	60	19	29	21
Age of firm in 1960:[2]						
10 years or less		8	7	0	4	3
11–20 years		4	4	0	2	2
21–40 years		13	12	0	9	5
41–60 years		20	16	7	6	7
More than 60 years		22	19	11	8	3
Unknown		16	2	1	0	1
Total		83	60	19	29	21

[1] Number of corporate income tax returns (for accounting year ending June 1961) classified by total assets. "Statistics of Income 1960–61," U.S. Treasury Department, Internal Revenue Service, p. 4.
[2] Years since incorporation in approximately present form.
SOURCE: Tables 10 and 11; "Standard and Poor's Corporation Records."

At least 60 per cent and usually 75 per cent of the output of the following sixteen metals is produced by firms with total assets of at least $250 million: antimony, arsenic, bismuth, calcium, columbium, gallium, indium, molybdenum, rhenium, selenium, sodium, tellurium, thallium, titanium (considering the size of parent firms), and zirconium. This list includes a large proportion of the metallurgical byproducts of major metals (particularly those of the large-tonnage subclass); it also includes the two milling coproducts produced by the Kroll process (titanium and zirconium), and the two metals without geologic limitations (calcium and sodium). If firms with assets of $50 million and above are considered, the following seven metals are added: cadmium, germanium, hafnium, scandium, thorium, vanadium, and yttrium, as well as many commodities not counted among primary products, such as individual rare-earth metals and most ferroalloys. The list then includes the remainder of the metallurgical byproducts and metals similar to them in their supply conditions, plus a few additional milling coproducts.

What, then, are the minor metals that are predominantly produced by "smaller" firms, say those with total assets of less than $50 million? Three of them, cobalt, platinum, and radium, can be put to one side because the small domestic firms are affiliated with large firms abroad. The others are: beryllium, cesium, lithium, mercury, rare-earth metals, rubidium, tantalum, and tungsten. These metals are all either individually mined or byproducts or coproducts of other minor metals; none is associated with a major metal. The structure for mercury and tungsten has always been strongly influenced by the mining of a large number of small ore deposits. Aside from these two, the metals produced by smaller firms are all newer metals and ones for which special production techniques have had to be developed. The list includes most of the metals recovered from natural brines and most of those recovered by solvent extraction or ion exchange. Annual production of these metals (or at least of certain of their primary forms, such as pure metal) is still small even in relation to the production of other minor metals. Typically they are produced by a preponderance of firms with assets between $5 million and $50 million, though in several cases American Potash, which has somewhat larger assets, is the largest single producer.

To a large extent the association of certain metals with these smaller firms can be attributed to the fact that only limited economies of scale can be attained. One reason why they are limited is that large amounts of source material do not have to be handled. Moreover, in many cases, economies that might be present at larger scales of production are not feasible given the current size of the market. As a result, small firms may be able to produce at the same cost as large firms. Not anticipating any advantage, the latter may simply stay out of the field at first. And once established in production, smaller firms may be able to remain profitable as a result of continuing research and their specialized experience.[12] However, when economies of scale are important at current levels of output, as with the byproducts of major metals or with the Kroll process metals, larger firms tend to play the leading roles.

Response to Increased Demand

Few new firms have grown up around the production of primary minor metals. Furthermore, it is the old and well-established firms in other fields, rather than the relatively new ones, that have been better able to step into production of minor metals. The bottom of Table 14 shows the age of firms producing minor metals. (Age is defined as the number of years between the date the firm was incorporated in approximately its present form and 1960. The age of the surviving firm was recorded in the case of mergers.) Of the firms for which information was available, about 30 per cent were incorporated before 1900, more than 50 per cent between 1900 and 1940, and less than 20 per cent since 1940. Of the firms formed since 1940, one is a joint venture of older firms, and a number of others had earlier histories under a different corporate organization.

Larger firms also tend to be older firms. Not one of the firms producing metallurgical byproducts of major metals is less than 40 years old. If demand for a metallurgical byproduct increases, the nonferrous metals industry responds in an orderly manner by expanding output through more complete recovery of the byproduct at old plants and the construction of new recovery facilities. It is very unusual for a young firm to commence production of a major-metal byproduct unless it can exploit sources not associated with a major metal. For example, the increased demand for selenium in the fifties was met almost entirely by nonferrous metals firms. However, several youthful firms were able to initiate production of vanadium as a milling byproduct of uranium because uranium was a new major metal.

Most of the minor metals producing firms that are under 40 years of age are either specialty chemical houses or firms specialized to the production of minor metals. Many of these firms are small but, regardless of their size, they are predominantly found producing such metals as cesium, lithium, rare-earth metals, and tantalum. These are all newer minor metals and their ores are generally available on world markets.

When demand for one of these metals increases, changes in the structure of production seem to follow inevitably. Often a number of new firms or

[12] The advantages of small firms are receiving some attention in trade journals. See *Chemical Week*, April 3, 1965, pp. 26–29.

joint ventures are created at the same time as existing firms are considering entering the industry. Commonly, a fair number of firms get beyond planning and into production; a competitive struggle ensues (delayed sometimes until completion of initial contracts); and after some years the industry settles down with fewer producers, most of which were well established before the new metal came onto the scene. Sometimes these structural changes take place through the exit of weaker (generally smaller or less integrated) producers; sometimes through merger or acquisition. Some minor metals have gone through this shake-out process; others, such as the refractory metals, are still in the midst of it; and for some it is no doubt yet to come. Although the response will be unique for each metal, the process can be illustrated by two brief case studies: sponge titanium and rare-earth products.

Sponge Titanium

Sponge titanium is elemental titanium metal that results from reduction of the tetrachloride by sodium or magnesium metal in the Kroll process. Although it must be melted to ingot (an additional refining step) before it can be formed and fabricated, sponge is generally considered as the primary product. The ore of titanium, rutile, is available in large quantities from several countries, and is traded on world markets. Titanium metal was first produced in 1948. Production grew rapidly during the middle fifties in response to a federal expansion program for the metal.[13] By 1957, six firms were actually producing sponge, several others had pilot plants, and one joint venture (Allied Chemical and Kennecott Copper) had announced plans for plant construction. None of the latter projects came to fruition because the federal demand for titanium metal was sharply curtailed in 1957, and recovery since then has been slow.

Of the original six sponge titanium producers, only two are left. Dow Chemical ceased production in 1957 though, as specified in its contract, it maintained its plant in standby condition until 1962; Cramet, a joint venture of Crane Company and Republic Steel, ceased production in 1958 and dissolved; Union Carbide left the field in 1962; and du Pont, the original producer, in 1963. Only TMCA, a joint venture of National Lead and Allegheny Ludlum Steel and the acknowledged price and production leader, managed to come through the lean years relatively unscathed. The other current producer, now called Reactive Metals, was formed as a joint venture by P. R. Mallory and Co., and Sharon Steel. Gradually, National Distillers

[13] The history of the titanium industry is extensively reported in the trade journals. In addition, see *The Titanium Metal Industry*, Report by the Attorney General under Section 708 (e) of the Defense Production Act of 1950, as Amended 1957; and Paul M. Tyler, *General Review of the Titanium Metal Industry*, Materials Advisory Board Report No. MAB–47–SM, 1958.

took over control of the concern as first Mallory and then Sharon Steel sold their interests; the concern has once again become a joint venture, this time between National Distillers and U.S. Steel. It is worth noting that both of the survivors are more fully integrated than any of the firms that dropped out. Each of them is integrated through all stages of production from mining forward, and they both sell sponge and ingot titanium to unintegrated firms.

Rare-Earth Products

Rare-earth metal products offer a second example of the changes in the structure of production that frequently follow growth in demand for a newer minor metal. This example contrasts with titanium in several respects: first, growth in demand was slower and steadier; second, the proportion of consumption taken by private commercial uses was much higher; and third, many of the firms involved are small relative to most others producing minor metals. Despite these differences, the gradual but substantial increase in demand has been accompanied by a similar reshuffling of the producing firms, as well as by numerous technical and marketing agreements between firms. Discussion of the agreements is deferred; here, the emphasis is on the reorganization of the firms.

One of the most important structural changes was the merger of older firms, partly or entirely specialized to rare-earth (and thorium) production, into larger and more diversified chemical concerns. Cases in point were the acquisition of Davison Chemical and Rare Earths, Inc. by W. R. Grace in 1954 and 1955, respectively; the merger of Lindsay Chemical (the largest single producer of primary rare-earth products) into American Potash in 1958; and the acquisition of Maywood Chemical by Stepan Chemical in 1959. The result was a considerable horizontal integration of the primary rare-earth industry.

During this same period, a large number of new firms throughout the United States entered business on the basis of production of rare earths. Among these firms were American Scandium, Lunex Company, Kleber Laboratories, St. Eloi Corp., and Research Chemicals (Division of Nuclear Corporation). Some older firms also initiated output of rare earths. Moly Corp., for instance, entered the industry after its discovery in 1952 of the world's largest known deposit of bastnasite (an unusual thorium-free rare-earth mineral). However, not all of the firms can be expected to survive, and some have not. Between 1960 and 1962 both St. Eloi Corp. and Kleber Labs went out of business; Stepan Chemical ceased supplying rare-earth compounds; and U.S. Semiconductor Products was bought out by Research Chemicals.

Other Metals

These two case studies could be extended by reference to other metals. The number of producers of cesium and rubidium grew from two in the early fifties to seven in 1962. The number of producers of primary columbium products grew from three in 1954 to eleven in 1958 and fourteen in 1962. That such growth alters the competitive setting is perhaps obvious; direct evidence is found in the fact that four firms produced 85 per cent of the columbium in 1958 but by 1961 seven firms shared this proportion of the output.[14] Moreover, in almost every case equally important structural changes have also taken place among unintegrated producers at later stages of production than those considered here. Using our case studies as examples, the surviving producers of misch metal (a mixture of rare-earth metals used in flints), of individual rare-earth metals, and of titanium ingot have been subject to the same mortality and reshuffling as have the primary producers. However, focusing on primary producers alone is sufficient to illustrate two points: first, that if the sources for these metals are available on the market, considerable changes in the structure of production commonly occur as demand expands; and second, that established and diversified firms have apparently been better able to cope with the technical and competitive problems of minor metal production than have new or youthful or specialized concerns. In fact, these responses are so typical that one must look for special conditions, as with beryllium, when they do not occur.

Inter-firm Associations

Specializations Within a Particular Minor Metal

One response to a situation in which markets for a particular minor metal are small is to carve out special markets within the over-all market for that minor metal. The goal of such a move is to find a sector of production with fewer competing firms or a sector that other firms can be forced or induced to stay out of. The specializing firm may gain some advantage by concentrating its research funds and its marketing activity on the products of this one sector. Specialization of production has sometimes been a way in which the smaller producers could maintain their position in a competitive market. Among other metals, this form of specialization is found with antimony, beryllium, rare-earth metals, titanium, and lithium.

In some cases the specialization is in no way based on the technology of production or the nature of source materials. Berylco has traditionally been the leader in the pricing and production of beryllium-copper alloys, but Brush, using an identical source material, has taken the lead in beryllium

[14] U.S. Bureau of Mines, *Minerals and Metals Commodity Data Summaries, op. cit.*

metal.[15] While there is a historical basis for this specialization, it is hard to understand why it should persist to the present. At the other extreme are cases in which the source material or technology dictates a form of specialization. For example, National Lead, the largest domestic producer of primary antimony, emphasizes antimony metal, but Asarco, the second largest producer, emphasizes antimonial lead. The reason is that National Lead has affiliates that mine antimony ore from which elemental metal can be recovered, whereas Asarco has affiliates that mine antimony-bearing lead ore that can be smelted directly into antimonial lead.

There can also be specialization to a particular market, rather than (or in addition to) specialization of product. The leading rare-earth processing firm, American Potash, produces for all markets but emphasizes the glass and ceramics industry. Michigan Chemical specializes in producing yttrium metal and compounds as catalysts; it also partially refines certain associated rare-earth metals and sells them to other firms for final purification. Moly Corp. concentrates both its research and its marketing activities on the use of rare-earth materials for steel additives. Similarly, Vitro specializes in the production of rare-earth hydroxides for the atomic energy industry, and Nuclear Corp. specializes in production for the aircraft industry.[16]

Specialization of production can be carried to a fine degree by tailoring production of a number of other products of the firm to the same market. American Potash markets not only its rare-earth products but also its boron products, its lithium products, and its cesium-rubidium products largely to the glass and ceramics industry.

Joint Ventures

Except with titanium, true joint ventures (50–50 sharing of investment and control) are rare in the history of minor metals production. And joint ventures in which one firm holds a majority interest and one or more others hold a minority interest are only a little more common. In either case, many companies that started as joint ventures were eventually bought up by one of the original partners, or, in a few cases, by an outside firm.

The advantages of joint ventures have been listed by Chairman Dixon of the Federal Trade Commission as follows:

To provide substantial capital to exploit raw material sources; to minimize the risk to one company in moving into a new industrial development;

[15] U.S. Attorney General, *The Beryllium Industry, op. cit.*, pp. 22, 44. *American Metal Market*, March 21, 1962, p. 14.

[16] See "Scanty Yield," *Barron's*, Vol. 34 (May 4, 1959), p. 11. Specialization is often dictated by factors other than the nature of the source material. For example, the emphasis of Moly Corp. on the steel industry is surely related to the fact that the company's main product has been ferroalloy metals; and it is likely that Nuclear Corp. is influenced by its location near aircraft plants in the southwestern United States.

to underwrite vast research programs, and to establish a single, joint facility which is more economical to operate than smaller, separate installations.[17]

For most minor metals, the first reason does not apply. The size of mining developments and their contribution to final value added are generally too small to induce association. Only when the mining operation involves a new and untried process do joint ventures come to be formed. A case in point was the exploration activity in western United States by the two major domestic beryllium producers. In each case, there was no proven method of recovering the beryllium-bearing mineral from the deposits, and the joint ventures ended when production from these deposits no longer seemed imminent.[18]

The other advantages listed by Chairman Dixon as underlying the formation of joint ventures remain valid. In fact, his analysis is given considerable strength by the fact that the joint ventures that are formed in the minor metals industry usually involve one of the newer minor metals. In addition there is one advantage that Chairman Dixon did not mention, that of associating with personnel who are already familiar with the area in which the operation is to be carried on. Most joint ventures in minor metals have involved association of an American firm and a foreign firm for production in the latter's country. Many of them, including the Kawecki-Durham joint venture to produce master alloys and refractory metals in England and the Fansteel-SGMH joint venture to produce refractory metals in Belgium, are set up with the local concern providing the management and the American firm providing technical personnel and equipment.[19]

What made titanium metal so different that a number of joint ventures were formed in the early years of the industry? A part of the answer is surely related to the fact that titanium was a case in which the normal caution about new products had to be tempered. Proponents of the metal's usefulness foresaw its transition from new metal to major metal within an exceptionally short time, and rather generous government contracts of several types were offered to aid this transition. However, there was only a small civilian market. (Even today military needs account for 85 per cent of titanium metal consumption.) And it was avowed government policy not to permit a monopolistic structure to develop in the new industry.[20] Thus, in contrast to most new metals, there was little possibility of innovator's profits and less assurance

[17] *The Wall Street Journal*, February 13, 1962, p. 2.

[18] Al Knoerr and Mike Eigo, "Beryllium Update–1961," *Engineering and Mining Journal*, Vol. 162 (September 1961), pp. 88–90, 93. Brush Beryllium acquired 100 per cent ownership of Beryllium Resources, Inc., from its partners in 1962; Berylco simply terminated its joint venture with United Technical Industries about the same time.

[19] *American Metal Market*, December 20, 1962, p. 1; also April 4, 1962, p. 14. *Metal Bulletin*, December 15, 1964, p. 21.

[20] Tyler, *Titanium Metal Industry, op. cit.*, p. 5.

of longer-run profits. In addition, it takes a combination of technologies to produce and fabricate titanium metal. The techniques for producing sponge titanium are more chemical than metallurgical, but metallurgical techniques (though not the common ones) are required to form the metal into ingots and to work it. The experience that National Lead had gained in working with titanium chemicals and titanium ferroalloys and that of Allegheny Ludlum in working with stainless steel were surely considered when they formed TMCA in 1950. The recent reorganization of Reactive Metals so that it is once again a joint venture, this time between National Distillers and U.S. Steel, derives perhaps from rather similar motives. It comes at a time when titanium consumption is recovering, but when major usage in either military airplanes or in supersonic commercial airliners remains quite uncertain.[21] And it combines National Distillers' experience to date and its production facilities (for sodium metal as well as for titanium) with the financial resources, research facilities, and distribution channels of U.S. Steel.

Joint ventures in which one firm retains a majority share seem to have a different cast than equally-owned joint ventures. The mutual benefit from the project presumably remains, but instead of the firms being joint producers, one is essentially a purchaser and the other a seller. The relationship amounts to a weak form of vertical integration. Typically, a joint venture is formed when a firm consuming some metal wishes to ensure its sources of supply or when a mine wishes to obtain some of the profit derived from the processing of its output. The former situation is illustrated by the 30 per cent interest held by Sylvania in Salt Lake Tungsten Division of Minerals Engineering; the latter by the 43 per cent interest held by Bikita Minerals, Ltd. of Southern Rhodesia in American Lithium Chemicals, a subsidiary of American Potash.

Exchange of Technical Information

The most common forms of agreement between two concerns producing minor metals are those for the purpose of exchanging research information and technical knowledge. Such agreements, which are common throughout the chemical industry, may or may not include the exchange of patent rights and mutual licensing. Like joint ventures, agreements of this kind are especially important between firms in different countries and most commonly involve the newer minor metals. The French firm Pechiney, for example, has been very active in forming agreements with American companies. Among

[21] See *Chemical Week*, March 28, 1964, p. 23; and *American Metal Market*, March 20, 1964, p. 1. The use of titanium was involved in the TFX controversy. The Boeing plane would have used titanium as a structural metal, but the General Dynamics plane uses stainless steel. *American Metal Market*, Aug. 19, 1963, p. 1. On the use of titanium in other new aircraft see *American Metal Market*, June 26, 1963, and *The New York Times*, March 3, 1964, p. 50.

others, it had an agreement with Grace on high-purity silicon and still has agreements with Sylvania involving certain refractory metals and with Vitro Corporation on rare-earth metals. The goal of the agreement with Sylvania was the broadening of both companies, for Pechiney has had more experience with columbium and tantalum, and Sylvania with tungsten and molybdenum. The agreement thus served to cut both the time and the initial investment necessary to develop a full research program in refractory metals. The 1960 agreement between Pechiney and Vitro ended their joint venture in Vitro Chemical (Vitro took over full ownership) but retained the technical agreements.

Domestic agreements for the exchange of patents and technical information are less common inasmuch as present interpretations of antitrust law place more severe restrictions on the patentees than when no exchange is involved.[22] ("Development association" or "information centers" can serve somewhat the same purposes, however.) Hence, the agreements are rarely made between firms producing similar products. More typical is one such as that between Michigan Chemical, a producer of rare-earth materials from ores, and Haveg Industries, an equipment designer and fabricator, to work mutually on the use of rare-earth metals in nuclear reactors. Again, such agreements serve as a weak form of vertical integration. If both firms are doing research, either one may discover information of use to the other and the result may well be to broaden the markets for both.

Associations for Marketing

In contrast with technical agreements, agreements for sales and distribution of certain products are common among domestic firms, particularly among firms producing a limited range of products. Among other things, such agreements give one firm an opportunity to specialize in producing for a single market that the other partner knows well. In return, the latter acquires a wider range of products to sell to this market. The newer metals are again most commonly involved for the obvious reason that they are not mature commodities and do not have regular channels of trade. For example, U.S. Borax and Chemical is the agent for Lithium Corporation and for Vitro to sell their respective products, lithium and rare-earth compounds, to the glass and ceramics industry. American Potash serves a similar function for Moly Corp., and U.S. Steel has recently followed up its move into the minor metals field via titanium with an agreement to handle distribution of the columbium and tantalum products of Kawecki Chemical.

Marketing agreements also provide the opportunity for one firm to avoid the problem of maintaining a sales force or developing sales contacts for

[22] *Hoffman's Antitrust Law and Techniques*, pp. 390, 392.

specialized products. For example, Englehard Industries, the major domestic fabricator and distributor of precious metals, is the sales agent for all platinum metals produced by Inco. It also markets the beryllium compounds produced by the Colorado concern, Mincon, thereby allowing that young firm to concentrate on developing its new production technique.

Summary

Several generalizations can be made about the firms producing primary minor metal products. First, the important producers of metallurgical byproducts of major metals are almost exclusively in the nonferrous metals industry. The production of most minor metals in other classes, however, is largely in the hands of firms in the chemical industry, broadly defined to include firms that have a large chemical division. Firms from other industries do not play a large role over all, though they may be significant in the production of a particular metal, as mining companies are significant in the production of mercury.

Second, firms producing primary minor metals tend to be large and to have diversified interests. Firms with total assets of less than about $50 million are important producers of some minor metals of smaller annual output (relative to other minor metals), but in general they are less common than larger firms, particularly in relation to the national size distribution of all firms. The few metals that smaller firms have been able to produce successfully are newer minor metals (or new forms of the metals) that are recovered from readily available source materials by such techniques as solvent extraction or ion exchange. In most cases, large and diversified firms have had an advantage either because large amounts of byproduct-bearing source material must be brought together or because large throughput is necessary to secure continuous or large-batch processing.

Third, the typical firm produces several minor metals rather than a single minor metal, and its output of minor metals is usually but a small proportion of its total operations. Figures defining this share are difficult to obtain. Union Carbide (a producer of a particularly broad range of minor metal products) stated that in 1961 all metals accounted for just 7 per cent of its sales volume.[23] Inasmuch as Union Carbide also produces a number of large tonnage ferroalloys and other metal products, minor metals themselves cannot be very important in establishing its corporate guidelines for pricing,

[23] *American Metal Market*, March 2, 1962, p. 14. Similarly, most of the metals production of National Distillers takes place in its USI Division, which in 1960 accounted for something under 19 per cent of the firm's total sales. *Chemical Week*, April 7, 1962, pp. 22–23. For a qualitative statement of this same situation, see Caleb Fay, "Why Metals Cannot Benefit under Increased Defense Appropriations," *The Magazine of Wall Street*, Vol. 106 (June 18, 1960), p. 374.

plant location, and other decisions. In short, the dominant interests of most firms producing significant quantities of minor metal products lie elsewhere. The exceptions are the few firms that specialize in beryllium, mercury, or to a much smaller extent the rare-earth metals, and some firms with assets under $50 million whose production of cesium, lithium, and a few other metals forms a substantial part of their activities.

Fourth, partial vertical integration is common, but complete integration is not. Metallurgical and milling byproducts of major metals are generally integrated from mining through metallurgical reduction and refining. In contrast, milling byproducts and coproducts of minor metals, and the metallurgical byproducts derived from them, are generally integrated from metallurgical reduction and refining forward to more highly processed products. The remaining metals divide about evenly between those integrated one way and those the other. There is a strong indication that integration plays an important role in company decisions about whether to produce primary minor metals.

Finally, the newer metals are often originally produced under highly uncertain and very fluid conditions. Often many more firms begin producing a metal than can remain in the field when the first burst of consumption enthusiasm is spent. It is among the newer metals that various forms of marketing agreements, patent pooling, and even joint ventures (otherwise rare with minor metals) are most commonly found. After a few years these industries settle down with fewer firms and a more orderly rate of growth. The producers that remain in the field are usually divisions of older and larger firms.

7

SOURCES OF COMPETITION

THUS FAR in Part II the structure of minor metals production has been treated in more or less descriptive terms in order to show how this branch of industry works. Here we use this information, together with the analysis of supply conditions from Part I, to show how well the structure is working. In this chapter the main sources of competition are indicated, and in the following chapter, where the price record for minor metal products is presented, we analyze the strength of these various forces. In both chapters the emphasis on primary products as opposed to source materials is retained.

The sources of competition in the production of minor metals can be divided into those associated with the entry of new producers and those associated with competition among existing producers. Entry is dealt with first because the opportunity, or lack of opportunity, for firms to commence producing minor metal products is perhaps the most important source of producer competition. This is followed by an analysis of other sources of competition—international trade, secondary production, etc. And, finally, consideration is given to the way producers view demand, for this, too, has an influence on competition.

Bases for Entry into Production of Minor Metals

As a rule, a firm decides to go into production of minor metals only when at least one of the following conditions holds: (1) it possesses adequate quantities of raw material containing the minor metal; (2) it has previous experience with similar products or a similar technology of production; or (3) it consumes the primary product in one of its operations. The first condition amounts to integration from source material forward to primary product, whereas the third amounts to integration from manufactured article backward to primary product; the second is diversification with respect to product but not with respect to process.

Entry by Integrating from Source Material to Primary Product

The fact that a firm owns or regularly handles a substantial quantity of a minor metal-bearing source material does not in itself prompt the recovery of minor metals, for gain or loss of control of the source material can be effected at different stages in the production process for the main product. Most critical as an entry-inducing force is control at the stage at which the immediate minor metal source (the byproduct proper) is separated from the main product. If the minor metal in question is a milling byproduct, control at the mill is important; if it is a metallurgical byproduct, control at the reduction and refining stage is important. In either case, integration may— and commonly does—extend back to the ownership of mines, but it is not necessary for inducing entry into production of the minor metal.

The influence of source materials as an entry-inducing factor is strongest among the byproduct classes. For example, hafnium is recovered exclusively by firms producing reactor-grade zirconium, indium by firms with zinc plants, and so forth. Some nonferrous metals firms have made extensive efforts to develop markets for minor metal constituents in their ore, as witness the research on rhenium by Kennecott and that on gallium by Alcoa. Nevertheless, possession of the immediate byproduct-bearing source material confers only an opportunity upon the firm. If a firm produces copper, and if it separates a selenium-bearing material at some stage of production, it can also produce selenium. But it need not do so. Generally speaking there are three alternatives. First, the firm can simply discard the minor metal-bearing material. Second, the firm can sell the minor metal-bearing material to another firm. Or third, it can produce the minor metal itself.[1]

So far as byproducts of major metals are concerned, there is seldom a competitive market in which the immediate byproduct-bearing source materials can be bought and sold, as are the major metal ores themselves. This might indicate that there is little difference between the economies of scale when saving source material and those when extracting a byproduct from it. This is not likely to be the case. The more likely reason for the absence of markets for byproduct-bearing materials is that the mining and extractive metallurgy of major metals is already concentrated. The firms that are large enough to recover, say, germanium or tellurium profitably are just those firms that possess the main product at the point at which the minor metal source is separated from it.

---

[1] See the section "Byproduct Production" in Chapter 4. There is one further alternative. A firm can have the minor metal recovered from source material by another firm on a toll basis. However, the production of minor metals by one firm for the account of another is not common. The only important case known to the author is the arrangement between Anaconda and Asarco under which the latter refines to metal the impure bismuth-bearing material produced by the former.

A second reason why markets for the immediate sources of major metal byproducts are not common is that the markets tend to be monopsonistic. This is the situation that faces the sellers of lead-cadmium residue. Separation of lead and cadmium is a necessary step in zinc smelting, and many firms go on to extract these two metals from the residue resulting from separation. However, some firms, generally smaller smelters, do not. But, Asarco is the only concern that has a "custom" plant to which such residues can be shipped. Its Globe refinery in Denver is unique in that it is operated chiefly to recover cadmium. The smaller plants must, therefore, either discard the residue or sell to Asarco.[2] The one other byproduct-bearing source material that is commonly sold is molybdenum-rich concentrate from copper mills; this finds a broader market because moly concentrate is also produced as a main product.

It is not surprising that large nonferrous metals firms dominate the production of byproducts of major metals. Not only do they possess source materials that cannot be purchased on a market and that they wish to utilize, but also, being integrated producers, they are able to take advantage of whatever economies of scale may be open. Other firms are practically precluded from entry because it is not ordinarily profitable to purchase major metal ores simply to obtain the small amount of minor metal that they contain.

For minor metals that are milling byproducts or coproducts of other minor metals the situation is quite different. Production of the immediate minor metal source is generally dissociated from production of the primary product derived from it, and source materials are sold in an open market. Because there are generally a number of purchasers and a number of uses for the sources of these metals, the mining and milling firms can market their output with little fear of facing a monopsony. At the same time, the numerous mines and the availability of marginal deposits of most milling byproducts and coproducts give assurance to producers of primary products that they will not be faced with a monopoly.

Although integration into mining—"going basic"—is not common, it has been an avowed goal of many primary producers. Under what conditions will a firm that is extracting a primary minor metal for any reason *other than* control of source material integrate backward into the recovery of source

[2] Custom smelters for major metals have been cited as an example of monopsony. Frederick T. Moore, "Industrial Organization in Non-Ferrous Metals," unpublished Ph.D. dissertation, University of California, Berkeley, 1951. Interestingly enough, some of the revolts that have occurred among miners against the nearest custom smelter have revolved around payment for byproducts. However, to break away from the nearest plant almost invariably involves considerable transportation cost to the miners. Hence, such revolts concerned payment for gold and silver, which have a much higher net value per unit of ore than do the byproducts under consideration here.

material? Two criteria seem to determine whether such integration will be seriously considered: (1) the primary product must be produced in rather substantial tonnage; and (2) there must be a reasonable possibility of a serious shortage of source materials. The latter condition means that integration will not be common among metals with heavy sand sources but will be spurred for tonnage metals with pegmatite sources. The only pegmatite metal that is produced in substantial amounts is lithium, and lithium production is integrated.

The influence of source materials as an entry-inducing factor among individually mined minor metals varies to extremes. Mercury mining is practically always integrated with metal production. Tungsten mining, on the other hand, is generally unintegrated. However, the same forces are at work on individually mined minor metals as on other classes of minor metals. First, ownership of deposits may spur production of primary products. Vitro Corporation of America, for example, once considered entry into the primary beryllium field. Although the firm was already engaged in the production of other minor metals and owned a mill which could be converted to the recovery of beryllium minerals, its interest in beryllium seemed to be largely predicated on the fact that it had discovered some beryllium deposits.

The case of beryllium exploration in western United States can also illustrate the second conclusion, that integration backward will be considered when large amounts of difficult-to-obtain source material are needed. Neither of the two major domestic beryllium firms is integrated into mining. The firms have emphasized that at the current level of production and with the adequate sources of supply for this level (even though these sources are foreign), they have no need to integrate into ore facilities. However, when in the early sixties interest in beryllium metal soared, both producers became active in exploration for and research on low-grade domestic ores. It is clear that had they needed large amounts of beryllium, production would have been obtained from domestic deposits, at least for leverage against foreign producers.

Entry by Diversification

The second condition that forms a basis for entry is that the technical experience of personnel in the concern be applicable to the minor metal in question. The experience could be with the same metal in a different form, but it is more likely to be with other metals in the same form. The similarities of properties and technology among minor metals in the same form are often more marked than are the similarities among different forms of the same metal. Union Carbide, for example, produces both ferrotungsten and ferrovanadium but contracts out for production of vanadium metal.

For the most part it does not appear that direct transference of technology has been a particularly successful basis for entry. Production of titanium pigments led both du Pont and National Lead into production of titanium metal, but there is little indication that their previous experience was of much value to them. (The experience of other firms with stainless steel was probably of more value.) Similarly, as the leading producer of germanium, Eagle-Picher also began to produce high-purity silicon, which is chemically similar to germanium and has the same electronic properties. However, in 1960 Eagle-Picher was forced to admit that it had been unsuccessful in its attempt to transfer germanium technology because of the greater difficulty of purifying silicon.[3]

One group of metals among which transference of technology has been successful includes the ten or so that are known as refractory metals. The most important are chromium, columbium, molybdenum, tantalum, tungsten, and vanadium. There has been a proliferation of refractory metals plants—some as joint ventures—around the country during the past several years.

Ferroalloys form another group whose production technology is transferable from metal to metal. Except for ferrophosphorus (which is produced at fertilizer plants) and ferromanganese (which is commonly produced by steel companies), ferroalloys are produced by firms that have an output mix ranging across several of them. Some firms have specialized in production of the ferroalloys of minor metals; Shieldalloy, for example, is one of four producers of ferrotitanium, one of six of ferrocolumbium, one of two of ferrotantalum-columbium, and one of three producers of ferrovanadium. Smaller companies can bridge this field because production of ferroalloys does not require such careful control as does production of metals. Iron, a common impurity, is of no consequence at all, and most source materials can be charged directly to the ferroalloy furnace.

Entry by Integrating from Manufacturing to Primary Product

The third condition that can induce entry is the need for the primary minor metal product in the process of manufacturing. This is the most important entry-inducing condition when source materials are not tightly controlled by other firms and when either relatively large quantities of the minor metal are regularly required, or the minor metal serves a particularly critical function in a continuing product line. For example, a number of tungsten carbide firms operate their own tungsten trioxide mills, and one or two have operated mines. Also, at one time, large producers of electric light

[3] "Chemical Processes Lead Race for Cheaper Hyperpure Silicon," *Chemical Week*, Vol. 87 (July 16, 1960), p. 64.

bulbs held interests in tungsten mines and trioxide plants, though most companies sold them when imported tungsten became much cheaper than domestic during the fifties. Similarly, several oil concerns have owned mercury mines from the time when it was a critical and expensive commodity in refining.[4] In other cases, integration does not go back as far as the primary metal. Firms such as General Electric produce a wide variety of metallic products, but do not as a rule enter into production of either ore or primary products. However, a few companies, including Fansteel and Sylvania, try as a matter of general policy to produce a large part of the primary products they require.[5]

Integration backward by consuming firms—commonly firms manufacturing chemicals—can come into conflict with integration forward by other firms if the latter attempt to secure the added profits from further processing of primary products.[6] For example, nonferrous metals firms, once content to produce germanium dioxide, now also reduce dioxide to metal. Similarly, intermetallics containing arsenic, tellurium, and bismuth are being produced by nonferrous metals firms as well as by a variety of other firms. Asarco and Cominco, in particular, seem to be emphasizing high-purity metals. If the past is a guide to the future, the competition for dominance of production stages between the recovery of source material and incorporation of the metal into manufactured components will be resolved metal by metal according to whether one group of firms can retain exclusive possession of the necessary source materials. If primary metals firms can do so, as with major metal byproducts such as germanium and vanadium, they will assume the added functions; if they cannot, as with tungsten and the rare-earth metals, production at the intermediate stages may well be incorporated by the manufacturing concerns that consume these metals.

Competition Among Existing Producers

The three alternative prior conditions for entry indicate that for some classes of metals there will be relatively few firms in a position to begin production, and that for other classes there will be broader possibilities for entry of new producers. Yet, as will be shown in the next chapter, the force of competition is not entirely circumscribed by the opportunities or lack of

[4] Among these concerns are Sun Oil (Cordero Mine), Colorado Oil and Gas (Abbott Mine), and Rare Metals Corp. of America (Idaho-Almaden Mine), which was organized by El Paso Natural Gas and Western Natural Gas.

[5] "Figuring into New Profits," *Chemical Week*, Vol. 89 (November 11, 1961), p. 47; "Status-Seeking Silicon," *Chemical Week*, Vol. 88 (April 1, 1961), p. 23. See also *American Metal Market*, July 18, 1961, p. 14.

[6] See, for example, "Status-Seeking Silicon," *loc. cit.; The New York Times*, June 16, 1960, p. 49, col. 1; and *The Journal of Commerce*, October 10, 1962, p. 3.

opportunities for entry. Nor is the picture completed by bringing in specific source materials, for resource appraisals indicate that sources for most minor metals occur rather widely throughout the world and appear to be adequate for expected levels of demand.[7] There are still other forces of competition, forces that do not depend so directly on the character of source materials, and it is to these that we now turn.

Concentration of Production

Firms producing primary minor metal products are generally few in number and large in size. In the case of only a few metals are there more than a half dozen producers, and commonly there are less than four (see Table 15). Moreover, it is common for one firm to supply 40 per cent or more of the annual domestic output. Price leadership usually accompanies such concentration. Thus, Asarco is the price leader for arsenic; Berylco for beryllium copper alloy; TMCA for titanium; and so forth.

Unlike production, the consumption of minor metals is typically divided among a large number of firms: bismuth is consumed largely by 60 firms; selenium is consumed by more than 200 firms; molybdenum by about 1,000; and cadmium by more than 1,500.[8] Even a relatively new metal such as elemental beryllium is consumed by about 75 firms.[9]

The combination of few producers and many consumers does not necessarily mean that there is little competition. Many chemicals are produced by only a single firm, but in most instances that firm cannot behave as a monopolist because of potential entry by other firms and potential competition from other products. Much the same is true for minor metals. The force of potential entry derives from the widespread existence of the three alternative prior conditions discussed above. Only a few nonferrous metals firms with the appropriate source materials or a few chemical firms that might consider integrating backward may well be enough to induce strong competition. Similarly, the possibilities both for intersubstitution among minor metals and for the use of other materials in place of minor metals grow with technology and sharply limit monopoly power.

Concentration in the production of minor metals has been increased by the tendency for larger firms to enter into the production of some minor metals through the purchase of specialty firms producing that product. Among the numerous cases that could be cited are the entry of American

[7] Hans H. Landsberg, Leonard L. Fischman, and Joseph L. Fisher, *Resources in America's Future* (Baltimore: The Johns Hopkins Press for Resources for the Future, 1963), pp. 447–52, 471–83.

[8] U.S. Bureau of Mines, *Minerals and Metals Commodity Data Summaries*, 1961.

[9] *American Metal Market*, November 20, 1962, p. 14.

Table 15. Estimated Concentration Ratios for Domestic Production of Selected Primary Minor Metal Products, 1960–61

Metal and form[1] (1)	Number of producing firms (2)	Concentration ratios		Firm(s) listed in col. 3 (5)
		Each column shows the number of firms producing a certain percentage of the output		
		(3)	(4)	
Antimony	9	1–35%	2–60%	National Lead[2]
Arsenic trioxide	2	1–70	2–100	Asarco
Beryllium				
Metal	2	1–75	2–100	Brush
Alloy	2	1–65	2–100	Berylco
Bismuth metal	3	1–90	2–98	Asarco[3]
Cadmium metal	11	2–60	5–80	Asarco and Anaconda[4]
Calcium metal	2	1–80	2–100	Pfizer
Cesium and rubidium compounds	5	2–80		American Potash and Dow
Columbium				
Metal and alloys	14	4–60	7–86	Wah Chang, du Pont, Union Carbide, Fansteel
Ferroalloy only	7	3–75		Union Carbide, Moly Corp., and VCA
Gallium metal	3	1–60		Alcoa
Germanium metal	4	1–40		Eagle-Picher
Lithium carbonate	4	2–80		American Potash and Foote
Mercury metal	75	6–85	20–98	
Molybdenum trioxide	7	1–70	2–85	Amax[5]
Rare-earth mixed compounds	10	5–90		American Potash, Grace, Vitro, Michigan Chemical, and Moly Corp.
Selenium metal	6	1–40	4–90	Asarco
Sodium metal	3	1–50		Ethyl Corp.
Tantalum metal	8	3–80		Fansteel, Kawecki, and National Research
Tellurium metal	5	1–40		Asarco
Thallium metal	1	1–100		Asarco
Thorium oxide	5	2–90		American Potash and Grace
Titanium sponge	4	1–50	2–75	TMCA
Vanadium pentoxide	7	2–75	4–98	Union Carbide and VCA

[1] When no concentration ratio could be found or estimated for a minor metal commodity, it is not listed. However, the bulk of the output of the following unlisted commodities is produced by the single firm indicated: indium (Asarco); rhenium (Kennecott); scandium oxide (Vitro); yttrium oxide (Vitro). Zirconium sponge production in 1960–61 was divided roughly equally among Pittsburgh Plate Glass, National Distillers (Reactive Metals), Wah Chang, and Carborundum Corp. Hafnium production was shared equally by the latter two concerns.

[2] Asarco is the second largest producer in total, but the largest producer of antimonial lead. Sunshine Mining is third largest with 7 or 8 per cent of annual production.

[3] USSR&M is the second largest producer.

[4] Top five producers include Amax, American Zinc, and New Jersey Zinc as well.

[5] Moly Corp. is the second largest producer (using molybdenum concentrates supplied by Kennecott Copper).

SOURCE: Bureau of Mines information; trade journal articles; corporation annual reports.

Potash and of Grace into the rare-earth industry, the entry of Anaconda into beryllium, and of National Distillers into titanium.[10] Alfred E. Kahn considers this type of expansion to be typical of the production of chemicals:

> A second evidence that size and fewness of sellers have not been the result simply of the economies of scale is the way in which chemical companies have chosen to expand. When branching into new fields, they have done so almost invariably in a manner best calculated to avoid trouble (competition); namely by collaborating with, or buying out, any large, established firms already in or planning to enter the field. . . .[11]

This manner of entry does not increase the number of producers, but it can make existing producers into stronger and more effective competitors.

International Trade

International trade has been an important force leading to keener competition in both minor metal source materials and primary minor metals. As discussed in Chapter 5, there is practically free trade in sources for minor metals. Admittedly, the free list includes many source materials for which there is little or no trade. But in those cases in which there has been direct competition between domestic and foreign source materials there is little question of the influence of the latter on domestic competition.[12]

Producers of primary minor metal products abroad must export to the United States over a tariff wall that for most products is between 10 per cent and 12½ per cent. A 12½ per cent tariff is a significant but not an insuperable barrier, and European and Japanese producers of many minor metals successfully compete with domestic producers. However, just as with potential competition from domestic firms, the *threat* of foreign competition is at least as important as actual imports. When establishing prices and marketing policies, domestic producers must take into consideration the conditions under which imports might begin to take over a significant part of the domestic market. For example, if U.S. producers cannot satisfy domestic demand at, say, the London price, imports are inevitable; if, as with cadmium, they can satisfy it, there is a strong incentive to keep their prices in line with this price.

[10] These mergers were discussed in Chapter 6.

[11] "The Chemicals Industry," *The Structure of American Industry: Some Case Studies,* Walter Adams, editor (New York: The Macmillan Co., 1950), pp. 205–6.

[12] See the Reports to Congress by the U.S. Tariff Commission under Section 332 of the Tariff Act of 1930; the subject is also discussed in the Reports of the U.S. Attorney General under Section 708 (e) of the Defense Production Act of 1950. See also Table 6.

Secondary Production

In general, the availability of scrap metal will induce greater competition. But the strength of the force varies from metal to metal. For some minor metals—notably antimony (especially antimonial lead), mercury, platinum, and the various forms of tungsten—secondary production is a major part of annual supply and one that responds fairly rapidly to price changes. For other minor metals it is a negligible influence because the metal is dissipated by the nature of its use, as with the cadmium used in paint or the thallium used as animal poison. For some others, such as beryllium and titanium, the market for scrap metal has been very limited because these metals tend to react with or absorb other elements and become contaminated. However, the situation is changing. It has been found that although slight contamination may render a metal unfit for the use for which it was originally produced, it does not necessarily render it unfit for other uses. Finally, the technology of secondary production is also changing so as to permit much cleaner and quicker recovery of the individual metals from scrap.[13] Thus, scrap minor metals can be expected to provide an increasingly important competitive force.

The availability of secondary metal may be reflected in lower prices. Scrap germanium, which is now recovered by most transistor manufacturers and returned to primary producers for re-refining, is given a good part of the credit for the lower germanium prices since 1957. In addition, new firms may enter an industry on the basis of secondary production. In 1963 Beryllium Metals and Chemicals (subsidiary of Lithium Corp.) became the third producer of beryllium metal with construction of a plant to electro-refine scrap to 99.5 per cent pure metal. The firm capitalized not only on a new process but also on the dissatisfaction of beryllium metal consumers with the refusal of the two established producers to buy back new scrap.

Government Programs

The effect of federal programs on the production of minor metals (to say nothing of major metals) has been so pervasive in the postwar period that it could not be adequately treated in a study of this scope or length. Rather, the role played by government has been introduced into the analysis whenever it was relevant to a question at hand. Here it remains to advance the

[13] See Arant H. Sherman, "Salvaging Hidden Values in Scrap" (text of an address at the 29th Convention of the Institute of Scrap Iron and Steel at Miami, Florida, on January 15, 1957); reprinted in *American Metal Market*, January 17, 1957. See also *American Metal Market*, December 19, 1962, p. 19; July 30, 1963, p. 16; January 8, 1965, Section 2, p. 18; and "Bismuth," by R. N. Spence, *Engineering and Mining Journal*, Vol. 164 (February 1963), pp. 126–27.

tentative judgment that on balance the results of the various federal programs have tended to increase competition. From the demand side, programs associated with defense, space, and atomic energy have both directly and indirectly boosted the consumption of many minor metals well above what it otherwise would have been. From the supply side, the federal government induced the creation of a substantial proportion of the production capacity that now exists for many primary products. Minor metals, of course, qualify for the various special benefits for minerals under our tax laws and loan programs, but the really important forms of federal aid have been rapid depreciation, "put" contracts, stockpiling (and stockpile upgrading) contracts, and other techniques for reducing, if not eliminating, the risk of investing in minor metals. The higher levels of consumption for older minor metals and the development of new minor metal products have served to break down barriers between the product lines of existing firms and to attract into production firms that had not previously produced minor metals. Competition has also been engendered by the availability of federal research funds. A concern with interest in some minor metal may be able to gain experience in production via research contracts for which the risk to the company is negligible. Even if the firm decides against entry, the increased potential competition can lead to more competition among the firms that are already producing the metal.

The conclusion that the federal programs leading to increased consumption and production of minor metals have strengthened competition should be qualified. By and large, they have strengthened competition among strong, established producers of some minor metal product (when the industry was previously in existence) or among established firms in similar fields (when the industry was totally new).[14] Moreover, the stronger competition does not usually appear until after the completion of government contracts.

Effect of Demand

Several characteristics of the demand for minor metals, as outlined briefly in Chapter 2, have important effects on the nature and force of competition. Producers commonly believe, or state that they believe, that the short-run demand for minor metals, particularly for those produced in relatively large amounts, is inelastic.[15] In a situation in which there are few producers, an inelastic demand will create a reluctance to initiate strong competition. Producers fear that price cutting will lead to a price war as each attempts to

[14] See the Reports of the U.S. Attorney General under Section 708 (e) of the Defense Production Act of 1950; especially *The Beryllium Industry*, 1957, pp. 41–44.

[15] *Supra*, pp. 34–35. See also *The Journal of Commerce*, March 28, 1962, p. 6; Kahn, *op. cit.*, pp. 209–11; *Metal Bulletin*, December 24, 1963, p. 19; and *American Metal Market*, February 10, 1964, p. 24.

gain a larger share of a market that is not enlarged by the price cuts.[16] This reluctance is reinforced by the uncertainty surrounding the demand for minor metals. Uncertainty arises both from rapid change in the consumption of products using minor metals and from substitution away from minor metals (or from one minor metal to another) in those products. The problem of profit maximization under conditions of uncertainty has been discussed in economics and business literature, and the broad conclusion is that the possibility of an unfavorable eventuality is typically weighted more heavily than is the possibility of a favorable one.[17] It has also been pointed out that when uncertainty is high, innovations may be slow to appear because of excessive charges for obsolescence.

The producer view of short-run demand, which tends to limit strong competition, can be offset at least in part by his view of long-run demand. If a producer feels that there is an opportunity to substantially increase the size of the market (the possibility that attracts most new entrants), he is often quite willing to engage in vigorous price or product competition. The opportunity may stem from the development of a really major market for the minor metal itself, as with the possible use of titanium for the structural metal in supersonic aircraft. Or it may stem from the fact that the minor metal is a small but critical part of another product on which profit hopes are pinned. The following answer was given to a query as to why companies are entering the columbium metal and alloy field: "Management is being convinced that expenditures in this field are essential, not because of the dollar value of the refractory metal business per se but because of the dollar value of the products using the refractory metals. We want to make certain that the performance of a few million dollars worth of refractory metals does not, in any way, jeopardize the performance of tens of millions of dollars worth of space systems and nuclear reactors."[18]

Thus, the production of primary minor metal products has been closely tied to applied research and technologic developments. Most firms producing primary minor metals are either chemical firms or nonferrous metals firms.

[16] The effect is similar to the reluctance to initiate strong competition that theoretically arises because of joint costs. Each producer fears that price will be driven down to the level of "out-of-pocket" costs, while the costs that cannot be allocated to a specific product will not be covered. But minor metal joint costs are rarely mentioned in this connection, and there is little evidence to indicate that they are an important factor limiting competition. One possible explanation is that for most minor metals the value added at joint stages of production is smaller than the specifically allocable costs at later stages, so that the spread between allocable costs and total costs is not large.

[17] William Fellner, *Competition Among the Few* (New York: Alfred A. Knopf, 1949), pp. 144–57, especially p. 150.

[18] D. T. Hurd of General Electric, quoted by Donald Peckner, "Columbium," *Materials in Design Engineering* (December 1961), p. 107. Dr. Hurd is manager of research in the Lamp Metal Components Department of General Electric.

The extent of the research effort by the former hardly needs to be documented. The nonferrous metals industry has lagged badly in providing funds for research, but its effort in this direction is growing and includes an important emphasis on the minor metals.[19] In short, technologic competition is common among firms producing minor metals. And, as elsewhere, technologic competition has proven to be a force that severely limits monopolistic power in both time and in scope.

[19] H. M. Bannerman, "The Search for Mineral Raw Materials," *Mining Engineering*, Vol. 9 (October 1957), pp. 1103–1108. D. Swan, "Industrial Research—Its Aims, Organization, and Facilities," in *Economics of the Mineral Industries*, Edward H. Robie, editor (New York: American Institute of Mining, Metallurgical, and Petroleum Engineers, 1959), pp. 655–76; *American Metal Market*, October 24, 1961, p. 14.

8

THE STRENGTH OF COMPETITION

HOW EFFECTIVE is competition in the production of minor metals? How efficient is the response of production to the forces of supply and demand and to the ever-changing results of their interaction? The most visible sign of the response is provided by the price record. However, prices alone do not tell us enough; they become meaningful only through comparison with detailed cost data. Cost data as such are not available. But the nature of minor metal production costs is implicit in the models of supply conditions and in the character of the production processes. Viewed against this background, prices are valuable, but not definitive, indicators of the state of competition.

The Price Records

In compiling price data, one would ideally like to know the actual prices at which transactions were completed. Such data are generally unavailable, and, even if they were available, collecting transactions data for 33 different metals would be an enormous task. Next best are quoted prices, which are readily obtainable. Over several years or more, the trend of list prices quoted by producers will parallel the trend of transactions prices and will reflect longer term supply-demand adjustments. Though in some cases it is difficult to find prices quoted for the same form and purity of a metal from year to year, the quotations are accurate enough for our purposes.

Information on quoted prices and price changes for the common forms of most minor metals is summarized in Table 16 for the years 1945 through 1961. Analysis of the data shows that most primary minor metal products exhibit one of four characteristic types of price behavior in this post-war period: a "hump" with high prices between periods of lower prices; a steady or nearly steady increase; a steady or nearly steady decrease; or fluctuating prices superimposed on longer term price trends. (For the most part price changes since January, 1962, do not alter the patterns significantly.) Of course, not every metal has a price record between 1945 and 1962 that corresponds to one of the four just described. Some, such as tantalum metal, show an independent course that cannot be dissociated from the changed purity and form of the metal. Nevertheless, most primary products follow one of the four patterns. Those that do are listed below:

Humped prices:

    antimony—exhibits more fluctuations than most others with this pattern[1]
    arsenic trioxide[1]
    cadmium—pattern ended in 1959 with price at 1947 levels; since then
        the price has increased sharply
    cobalt
    columbium ferroalloys
    germanium[2]
    misch metal (rare-earth alloy)
    selenium[2]
    tantalum ferroalloys
    thorium—stable price for nitrate and alloy-grade metal since 1952
    thallium—price returned to 1945 level before Korean War

Steady or nearly steady increase:

    beryllium in copper alloy[2]
    bismuth—stable price 1950 to 1964[2]
    calcium metal
    molybdenum trioxide and molybdenum concentrate[2]
    radium—stable price since 1946
    sodium metal—one decrease
    tellurium (?)—may eventually follow the pattern of selenium
    vanadium pentoxide—two small decreases

Steady or nearly steady decrease:

    cesium compounds
    columbium metal
    gallium—virtually stable price 1945 to 1961[2]
    indium—stable price 1946 to 1958[1]
    lithium carbonate—one small increase[2]
    rare-earth (individual) compounds—quoted since 1956 only
    rhenium
    rubidium compounds and metal
    titanium sponge[2]
    zirconium sponge (commercial-grade and reactor-grade)

Fluctuating prices superimposed on longer swings in price:
   mercury[1]
   platinum-group metals—most platinum-group metals exhibit a broad
      humped pattern underlying the fluctuations; the price of palladium
      has been nearly steady over this period.
   tungsten trioxide
   sources for most minor metals

   [1] Pattern broken by price increase(s) since 1962; in the case of metals with fluctuating prices, it is only the pattern of longer swings that has changed.

   [2] Pattern continued by price change(s) (up or down as appropriate) since 1962.

The postwar hump, illustrated for four metals in Figure 5, is exhibited most characteristically by metallurgical byproducts of major metals, but also by some ferroalloys of minor metals. In most instances, prices began to rise sharply in 1946 when controls were removed and continued to rise until a year or two after the Korean War, or a little longer if a stockpiling program was in effect; after that, prices decreased, commonly to about the 1950 level, but in some cases even further. A few metals have earlier humps, as does thallium, and several have later humps, as do germanium and selenium, depending on the specific supply-demand situation.

The second and third types of price record are opposite trends; a steady or nearly steady increase since 1945 and a steady or nearly steady decrease since 1945 (Figures 6 and 7, respectively). The increasing price trend is found among a group of metals that not only cuts across minor metal class lines, but includes both older and newer minor metals and both large tonnage and small tonnage ones. For some metals, the increase has been moderate, lagging behind the commodity price index; for others, it has outdistanced the index. (The rise in the commodity price index over the period under consideration has the effect of deflating price decreases and exaggerating price increases.)

More primary minor metal products have exhibited a declining price trend than a rising one. In addition, none of the increases was as sharp as some of the decreases: gallium cost $1,350 per pound in 1946 and was more or less steady in price until 1961 when it fell to around $900; indium in the middle forties fell from $146 per pound to about $33, and then in 1958 fell further to $18; and prices of individual rare-earth compounds dropped 50 to 80 per cent between 1956 and 1959. Most metals that show such striking price cuts are newer minor metals for which the product itself changes so rapidly that it is difficult to make price comparisons over time. Indeed, in numerous instances the lower price became effective when a higher purity product was put onto the market. More significant are the smaller but more frequent price cuts that have occurred among larger tonnage products, such as lithium carbonate and titanium. The annual value to consumers of such cuts is greater even though the absolute decrease is less than with the small tonnage products.

Each of the three preceding types of price record is characterized by periods of price stability before and after price changes. (Note the relatively few changes in the quoted price for most minor metals listed in Table 16.) The fourth and final type of postwar price record found among minor metals is one exhibiting many fluctuations (usually of small amplitude individually) superimposed on some longer swings in price. Only a few primary metals, notably mercury, tungsten trioxide, and the platinum-group metals, exhibit

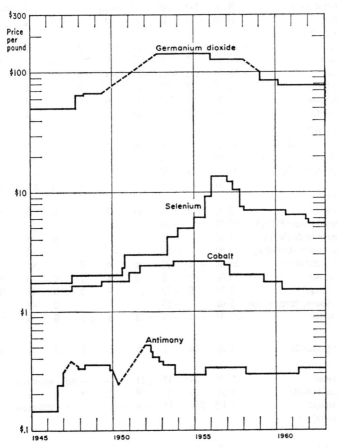

Figure 5. Selected examples of primary products that exhibit a humped price record, 1945–62.

Table 16.   Changes in Quoted Domestic Prices for Selected Primary Minor Metal Products,
January 1, 1945 to December 31, 1961

Metal and form by class	Price per lb. Jan. 1, 1962[1]	Percentage changes in quoted price			Number of times the quoted price was changed					
		1945–52	1953–61	1945–61	1945–52		1953–61		1945–61	
					Up	Down	Up	Down	Up	Down
Class A-1:										
Antimony metal	$0.325	138%	−6%	124%	>5	>8	2	3	>7	>11
Arsenic trioxide	0.04	37	−27	0	5	3	0	1	5	4
Bismuth metal	2.25	80	0	80	5	0	0	0	5	0
Cadmium metal	1.60	122	−20	78	8	2	4	4	12	6
Cobalt metal	1.50	60	−37	0	4	0	1	4	5	4
Gallium metal[2]	660.00	0	−48	−48	1	1	0	>2	1	>3
Germanium dioxide	76.00	184	−46	57	many	0	1	>4	many	>4
Indium metal[3]	18.25	−70	−44	−83	0	3	0	2	0	5
Platinum-group:										
Iridium	1,020.00	12	−62	−58	>6	>13	3	8	>9	>21
Palladium	350.00	0	0	0	0	0	>4	8	>4	8
Platinum	1,200.00	166	−12	134	>9	>8	>5	>15	>14	>23
Selenium metal[2]	5.75	72	92	228	3	0	5	5	8	5
Tellurium metal[2]	6.00	0	243	243	0	0	9	1	9	1
Thallium metal	7.50	0	−40	−40	2	2	0	1	2	3
Classes A-2 and B-2:										
Beryllium, 4% Cu alloy	43.00	120	30	186	5	1	3	0	8	1
Cesium chloride (technical)	27.50		−60					>2		
Columbium, metal sheet[2]	45.00	−50	−60	−80	0	2		>1		>3
Lithium carbonate	0.67	−36	−20	−46	0	>4	>1	>2	>1	>6
Rare-earths:										
Misch metal	2.70	0	−40	−40	3	1	0	5	3	6
Lanthanum metal (99.9%)	[4]152.00		−30				0	2		
Lanthanum oxide (98%)	[4]7.60		−45				0	4		
Rubidium chloride (technical)	13.00		−80					>2		
Tantalum, metal sheet[2]	55.00	−35	31	−15	0	2		>1		>3
Thorium nitrate	3.00	50	0	0	4	1	0	0	4	1
Titanium sponge[2]	1.37	0	−73	−73	0	0	0	15	0	15
Zirconium sponge (commercial)	5.00		−50				0	>2		
Rhenium metal[2]	600.00	−83	−25	−78	0	>1	0	3	0	>4
Class B-1:										
Vanadium pentoxide	1.38	16	8	25	3	2	2	1	5	3
Class C:										
Mercury metal[5]	2.50	39	−12	21	——————not applicable————————					
Molybdenum trioxide	1.59	43	39	99	2	0	5	0	7	0
Tungsten trioxide (concentrate)[2,5]	1.125	267	−65	−6	——————not applicable————————					
Class D:										
Calcium metal	2.05	11	0	0	2	0	0	0	2	0
Sodium metal	0.17	10	3	13	1	0	1	1	2	1

NOTE: To the greatest extent possible the calculations used in preparing the price data for each metal refer to a primary product of like purity, form, shape, etc. In a few cases adjustments have been made in quoted prices to obtain a price more comparable to one quoted at a different date. Gaps in the table indicate that the commodity was not produced at the beginning of the time period shown in the column heading.

No attempt has been made to correct prices for inflation. The Wholesale Price Index of the Bureau of Labor Statistics, U.S. Department of Labor, was at the following levels in the months for which prices are quoted:

All commodities:
(1947–49 = 100)

January, 1945	68.2
January, 1953	109.9
January, 1962	119.7

If it is assumed that mining and metallurgy use an "average" input of goods and services, the prices in Table 16 could be deflated with the result that any price increases would be relatively reduced and any price reductions relatively increased. The percentage changes in the Wholesale Price Index corresponding to the points in time compared in Table 16 are as follows:

January, 1945–January, 1953	+61%
January, 1953–January, 1962	+ 9%
January, 1945–January, 1962	+75%

[1] Prices represent the lower limit of large-lot prices for commercial-grade product, f.o.b. producer's plant, unless otherwise specified.
[2] Purity of the commodity increased somewhat over the time period for which information is given.
[3] Price steady from the end of 1945 to January 1, 1953.
[4] Price changes date from 1957.
[5] New York spot prices given; prices fluctuate daily.

SOURCE: "Minerals Yearbook" and "American Metal Market."

this pattern. However, it is typical of ore sources for minor metals. In general, the longer swings in price parallel the pattern of metals exhibiting a humped price record (except that prices fell from 1945 to 1950 in some cases), suggesting that the underlying economic forces may have been similar.

Minor Metal Commodities That Have Exhibited Fluctuating Prices

With the exception of the platinum metals, the commodities whose prices have fluctuated since 1945 are all products for which production rates can be easily altered. They are either mine products or primary products that do not require extensive processing at the metallurgical reduction and refining stage. Among primary products, the pattern is most common for those produced wholly or partly in the individually mined class. These two generalizations are supported by the record for antimony, which exhibits a considerable number of price changes superimposed on a humped price pattern. Although antimony was classed as a metallurgical byproduct, a substantial part of its annual supply is derived from direct smelting of individually mined ore, and an even larger part from scrap.

In most respects, the minor metals that have exhibited fluctuating prices are produced and sold under conditions approaching those of pure competition. Typically they are sold in highly responsive commodity markets with

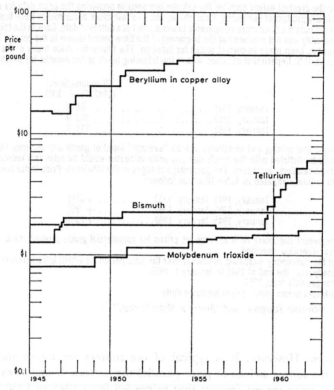

Figure 6. Selected examples of primary products that exhibit a
rising price record, 1945–62.

both spot and future contracts available. Moreover, again with the exception
of the platinum metals, there are so many producers in this country (actual
and potential) that monopolistic control would be difficult to effect. Finally,
foreign competition is extremely important.

A foreign industry, even when it is highly concentrated, can create
competitive conditions in the domestic industry. So long as substantial
imports are required to satisfy domestic demand, and so long as they can be
obtained at published delivered prices, domestic producers will be forced to
act as if they were in a purely competitive market. They must meet this
established price, but they have the opportunity to sell all they can at the
price.[3] Domestic mercury and platinum producers operate under such

[3] The effect can be shown by the dominant firm model. See fn. 9 below. The dominant
firm represents the conditions established in a monopolistic world market, and the smaller
firms represent conditions facing domestic producers.

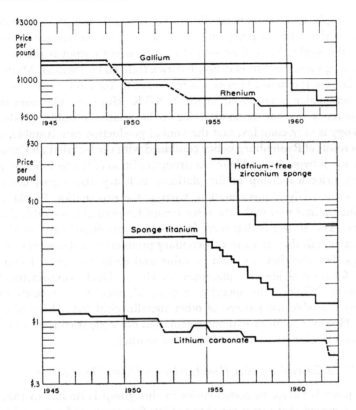

Figure 7. Selected examples of primary products that exhibit a falling price record, 1945–62.

competitive conditions. In the case of mercury, a few mines in Italy and Spain (mostly government owned or supported) dominate markets in the Western world. They have been strong enough to form cartels in the past, and they can still pretty well set price by entering or withdrawing from the market.[4] The 50 to 100 domestic producers, who generally account for less than half of domestic consumption, can do little but meet their price, whatever it is.

[4] See James W. Pennington, *Mercury, A Materials Survey*, U.S. Bureau of Mines Information Circular 7941 (Washington, D.C.: Government Printing Office, 1959), pp. 2, 12, 77; U.S. Tariff Commission, *Mercury (Quicksilver)*, Report on Investigation No. 332–32 (Washington, D.C.: Government Printing Office, 1958), pp. 44–53, 60–71; *Metal Bulletin*, January 8, 1963, p. 20; and February 7, 1964, p. 24.

The supply and demand conditions for the platinum-group metals are similar to those for mercury in that there are few important producers around the world. More than 40 per cent of world output is produced by Rustenburg Platinum Mines in South Africa (refined by Johnson, Matthey & Co. in London); almost 30 per cent is recovered by Inco in Canada; and most of the remainder is produced in the U.S.S.R. However, platinum differs from mercury in that demand seems to be relatively stable, production technology is very complex, and the annual production rate is stable. (This last is a result of Rustenburg policy combined with the fact that all Canadian production is byproduct.) There is a strong tradition of Rustenburg-Johnson, Matthey price leadership in the platinum industry and a past history of cartel action.[5] Given this situation, what is the explanation for the rapid price fluctuations that make the price record for platinum resemble that for mercury? The answer is that there is a fairly competitive "free" market for platinum metals that arises from secondary production and from speculative holdings. Because they are high in value and easily transported either in refined form or as jewelry, platinum metals are ideal "commodities" for speculation. Without the speculative demand, their price records would probably resemble the pattern of other metallurgical byproducts of major metals. However, with or without speculation domestic producers will be in a situation similar to a purely competitive market.

Minor Metal Commodities That Have Exhibited a Humped Price Record

The price behavior of commodities in this group is similar to that just described except that prices are changed less frequently and are consistently upward for part of the period and then consistently downward. The humped pattern is most clearly shown by metallurgical byproducts of major metals. Prices of almost every important byproduct metal increased as a result of a tight supply situation just after World War II or during and after the Korean War; they leveled off for a few years and then slumped in the late fifties when the situation changed to one of eased demand plus expanded major metal production. Selenium and germanium offer good examples of these effects (Figure 5). Each had a stable price until demand for the metal began to grow rapidly; prices then rose, and later fell after demand had relaxed.

[5] On current pricing practices see *Metal Bulletin* (London), February 27, 1959, p. 20; January 26, 1960, p. 24; and August 14, 1962, p. 20. See also *Minerals Yearbook*, 1959, p. 855. On cartel action see Gar A. Roush, *Strategic Mineral Supplies* (New York: McGraw-Hill Book Co., 1939), pp. 331–36. Rustenburg has also kept much of its potential output of less common platinum-group metals off the market in order to support the price. This practice can be followed because byproduct sources, such as those of Inco, are high in palladium but contain little osmium, rhodium, or ruthenium. Because the proportion of palladium in Canadian sources is so large, palladium is relatively cheap here (compared with other platinum metals). Also, Englehard Industries, the firm that markets Inco's output, is the price leader rather than Johnson, Matthey.

Tellurium, sister metal of selenium, may eventually follow the same pattern; its price rose rapidly between 1958 and 1961, then leveled off, but has yet to begin falling (Figure 6). Thus, there is nothing in the hump itself that necessarily reflects anything more than the play of supply and demand in a situation of naturally limited supply.

However, for some commodities exhibiting a postwar hump it is probable that limitations on entry permitted prices to be maintained at a high level for several years after the supply had become adequate to meet demand at a lower price. The price of selenium was nearly stable between 1957 and 1962, for example, despite a significant decrease in demand.[6] Similarly, the price of tellurium has remained at $6 per pound for the past four years despite a drop in apparent consumption. Moreover, whenever demand presses toward the limit of available supply, producers tend to satisfy the demand for high-purity metal first. High-purity grades of metal generally sell at a considerable price premium over commercial-grade metal. The common technique is to raise the price of commercial-grade metal while holding the price of high-purity metal constant and, if necessary, allocating the output of commercial-grade metal. This is possible only for metals such as selenium and tellurium for which entry is difficult.[7]

Nevertheless, although substantial monopoly profits on metallurgical byproducts of major metals are possible in the short run, they are limited in the long run by potential entry, by foreign competition, and by secondary recovery. For example, the number of domestic producers of germanium increased from one to four during the period of rising prices with the result that primary production grew rapidly; secondary production, an entirely new field, also grew rapidly with the entry of two firms. In most instances, foreign production has climbed by an even greater proportion than domestic, largely because byproduct supplies have been less fully exploited abroad. But for some metals monopoly profits may persist even in the long run because

[6] See *Minerals Yearbook*, 1960, p. 1257. See also the oblique references to this point in *Metal Bulletin*, August 23, 1960, p. 24; November 3, 1961, p. 26; and November 2, 1962, p. 21. The 1961 article followed a statement by American producers about selenium availability "to counteract rumours of a continuing shortage." Noting that "for months now, the selenium market in Europe has been as weak as water," the article went on to say, "the inappropriate nature of the latest statement about selenium has further weakened market confidence here in the value of statements made by these producers."

[7] Someone in the tellurium industry made the statement to a reporter that producers were following the principle of "dumping the bird in the hand to get at the two [thermo-electrics and free-machining steel] in the bush" (*New York Times*, June 16, 1960, p. 45). The article went on to point out the price differential involved. *Metal Bulletin* picked up on the analogy and said that it was "an unusual reversal of the well worn principle" (May 25, 1962, p. 20). They noted that, "Although the tellurium supply position is normally limited . . . the current practice of restricting sales has artificially worsened it. . . . It is obviously in the interest of producers to promote this new and high-priced usage for their material even if it is partly at the expense of established consumers."

supply expansion is limited by a scarcity of source materials; there are no new tellurium plants under construction in this country in spite of the price incentive.

In short, depending on how closely demand is pressing on available sources of supply and on how many firms are producing the given metal, the conditions of competition for metals exhibiting a humped price record between 1945 and 1962 can vary from those of pure competition to those of a loose oligopoly. When there are more than about five important producing firms or when there are opportunities for expanding byproduct supply or for bringing in supplies of a different character, prices appear to change fairly regularly (say once or twice a year) in response to changes on both domestic and world markets. If the number of firms is still larger, or supply limitations looser (say because of secondary production), prices begin to fluctuate like those of the metals discussed in the preceding section. On the other hand, when there are fewer producers, and when the possibilities for increasing the supply of the byproduct without increasing the supply of the major metal are limited (as they typically are for most of the large tonnage byproducts, such as cadmium), price leadership is common and downward price adjustments are relatively infrequent compared with upward revisions. In either case, however, substantial profits can be earned during periods of expanding demand because the rate of byproduct production cannot be rapidly increased beyond a certain point.

Minor Metal Commodities That Have Exhibited Rising Prices

Metals that have shown steadily increasing prices between 1945 and 1962 are at the opposite end of the competitive spectrum (so far as the domestic market is concerned) from those for which prices have fluctuated. Eight metals have exhibited rising prices since 1945: beryllium, bismuth, calcium, molybdenum, radium, sodium, tellurium, and vanadium. Though a rising price does not in itself indicate monopoly power, there is additional evidence in most of these cases. For instance, long periods of price stability seem to be associated with the pattern of rising prices, and not just because consumption rates are very low as with some of the newer minor metals. The price records for bismuth and tellurium are typical in that a series of price increases occurred in a relatively short period of time and followed, or were followed by, many years of unchanged prices.

Little information is available on pricing or competition in the production of either calcium or sodium. With calcium there appears to be a large measure of self-supply by ferroalloy producers; nevertheless, imports of calcium metal from Canada probably set the market price. Sodium capacity has expanded considerably in recent years while consumption has reportedly lagged, and there have been hints that a drop in the price, which has been stable since 1957, would be in order.[8] The fact that there are only three domestic sodium producers—Ethyl Corp., du Pont, and National Distillers, all large chemical firms—may explain why price has not dropped. There is no international trade in sodium metal.

The sharpest price increases have been shown by beryllium in copper alloy, which by 1962 had increased 186 per cent over 1945 and 30 per cent over 1953, and by molybdenum trioxide, which by 1962 had increased 99 per cent over 1945 and 39 per cent over 1953. The beryllium and molybdenum industries can usefully be contrasted because the source of monopoly power is different in the two cases.

Amax produces 60 per cent of the world's mine production of molybdenum from a single large ore deposit. Its only important competitors produce molybdenite as a byproduct of milling copper ore, and Amax acts toward these competitors exactly as would be expected of a dominant firm.[9] Amax sets the price for long periods of time and regulates its production according to the current supply and demand situation. The output of molybdenite from copper firms is determined by their output of copper (which determines the long-run potential supply of byproduct molybdenite) and by the Amax price (which determines the recovery rate for byproduct molybdenite, or the short-run supply).[10] As a further line of defense, the molybdenum industry in this country is protected by tariffs on both ore and metal products. Finally, the price position of Amax has been immeasurably strengthened by the steadily increasing consumption of molybdenum around the world (a technologic trend that Amax's own research effort has done much to further).

In contrast with molybdenum, the source of monopoly power in the production of beryllium cannot stem from control over raw materials as there is no shortage of good quality beryl on the world market. Since 1953, the

[8] "Who'll Use All That Sodium?" *Chemical Engineering*, Vol. 66 (November 2, 1959), p. 40; and "Sodium," *Chemical Economics Newsletter*, Stanford Research Institute (May-June, 1962).

[9] The behavior of a dominant firm is predicted by a model that illustrates the situation in which an industry is composed of a leading firm, which supplies at least 25 per cent of the industry output (a figure exceeded by the leading producers of many minor metals), plus a number of smaller firms. As shown by Professor Stigler, the dominant firm and the smaller firms will act differently under these circumstances. The former will maximize profit by pricing as if it were a monopoly facing not the total market demand, but a demand schedule that excludes the amount that the smaller firms can produce at each price level. The smaller firms accept the price set by the dominant firm and maximize profits by reacting to it as they would in a purely competitive market. The effect is that their production is just sufficient (when added to the production of the dominant firm) to satisfy the total quantity demanded by the market at that price. George J. Stigler, *The Theory of Price* (New York: The Macmillan Co., 1949), p. 227.

[10] *Metal Bulletin*, April 14, 1960, p. 23; January 11, 1963, p. 16; May 15, 1964, p. 25; and *American Metal Market*, April 3, 1964, p. 24.

price of imported beryl has decreased by almost one-third. However, the price of beryllium-copper continued to increase into 1955 and then remained steady until it increased again in 1962. The discrepancy is partially explained by the tariff situation: ores are admitted free but there is a 22½ per cent duty on alloys. More important is the absolute cost advantage of the two important producers, Brush and Berylco, through patent control and research. In addition, both firms deliberately maintain a level of production capacity for most beryllium products that is usually well in excess of their sales. Although, as noted above, there is both consumer dissatisfaction with some of the policies of Brush and Berylco and interest in new production processes, no one has yet challenged the two established firms in producing the main primary beryllium products.

Bismuth, like most other metals showing a rising price between 1945 and 1962, is closer to the molybdenum case than to the beryllium.[11] There are only about ten substantial producers of bismuth in the world; all are associated with the nonferrous metals industry. In 1931 Paul M. Tyler indicated that some attempts to increase sales by lowering the price had failed (he did not say for how long the price had been lowered), and that producers "deemed it wiser to market a smaller quantity at a remunerative price rather than to hazard their present fairly safe profits against the doubtful prospect [of higher volume production]."[12] The price of bismuth remained in one of its characteristic periods of quiescence from 1950 to 1964, and over that interval there was no need to question Tyler's opinion. From early 1964 to 1965 the bismuth price nearly doubled in response to new markets for the metal (mostly as a catalyst in the manufacture of acrylic fibers). In this instance, producers apparently felt so confident of the higher level of demand that they did not fear the loss of some traditional markets. But, in part, the higher prices are the legacy of tight producer control over bismuth which has further limited the naturally inelastic byproduct supply situation.

The structure of production for radium is an exaggerated version of that for bismuth. Since 1922 the dominant world producer has been Union Minière, which recovers radium-bearing slimes from uranium ore processing in the Congo and ships the slimes to an affiliate in Belgium for refining. In

[11] Although it has not been considered as a minor metal, the price record of nickel is remarkably like that of molybdenum—a steady and substantial rise since 1945. The competitive conditions are also similar. Inco produces around 60 per cent of the world supply but also engages in extensive market development. However, the recent slowdown in the rise of nickel consumption, combined with the new entrants into the field and with Inco's own expansion, finally resulted in a price decrease initiated by Falconbridge, the second largest producer.

[12] Paul M. Tyler, *Bismuth*, U.S. Bureau of Mines Information Circular 6466 (Washington, D.C.: Government Printing Office, 1931), p. 2. At the time Tyler was writing, an informal bismuth cartel was in operation.

the thirties a cartel was formed by Union Minière and Eldorado Mining & Refining of Canada, the only other important producer. Today the Canadian mine is closed and Union Minière's competition comes not from other producers but from artificial radioisotopes. (The medical use of natural radium has been officially discouraged in recent years because of a high contamination hazard not found with substitutes.) Nevertheless, in spite of the fact that production capacity is known to exceed consumption, the price of radium has not changed since 1946.

The case of vanadium is somewhat different in that it involves monopsonistic control over raw materials by two primary producers at the milling stage. In 1959 a group of mining and milling interests won a $4.5 million antitrust decision against Union Carbide and Vanadium Corporation of America in the U.S. District Court in Salt Lake City. They had charged the two companies with a conspiracy to control markets and prices of vanadium ore, concentrates, and ferrovanadium, as well as with nonpayment to vanadium mines for the uranium recovered and produced as a byproduct.[13] In 1957 the Justice Department had lost a suit which accused the companies of controlling vanadium pentoxide production and refusing to sell oxide to ferrovanadium producers. It is interesting to note in this respect that whereas price changes for molybdenum products and molybdenum ore (both of which are marketed by Amax) have been made simultaneously, the vanadium pentoxide producers (who are only primary product producers, not miners) have increased the prices of their vanadium products several times while being able to hold the contractual payment for purchases of vanadium ores to a single increase over this period.[14]

Minor Metal Commodities That Have Exhibited Falling Prices

Little can be said about competitive conditions for metals that have exhibited rather steadily falling price trends since 1945 except that technologic competition outweighs other aspects of competition and overshadows raw material supply limitations. This group includes the bulk of the primary products that are milling byproducts and coproducts of other minor metals and also most of the small-output metallurgical byproducts.

There are a number of reasons for the importance of technologic competition. The environment in which the firms produce is often surrounded with uncertainty. Few markets are even moderately safe or free from substitution.

[13] *Chemical Week*, Vol. 58 (December 26, 1959), p. 23. In another case, which involved Canadian production, the Supreme Court upheld the right of Continental Ore Co. to sue for antitrust damages from Union Carbide and VCA (*American Metal Market*, June 26, 1962, p. 1).

[14] Bureau of Mines, *Mineral Facts and Problems*, Bulletin 585, Washington, D.C., 1960, pp. 946–47.

Much pricing is experimental. One firm recently stated that it was cutting the price of a new metal by a substantial amount because a market survey had indicated that some people might be interested in it if the price were cut in half. Long periods of price stability may ensue for these metals so long as consumption rates are low and their technology remains static. But if new products are being developed and prospective markets are opening up, and if producers are engaged in research on extraction and preparation technology, the competitive situation can change quite rapidly. Price cutting and product improvement are among the few alternatives in this situation, and producers *must*, therefore, pass along the results of technologic progress. In the cases of cesium, rubidium, and most of the refractory metals, competition has been intensified by the existence of more firms relative to annual output than is typical of primary minor metals. The so-called overpopulation and resulting strong competition in the refractory metals industry is the subject of frequent producer complaints. Finally, international trade commonly expands as soon as there is a tonnage market for these products.

Summary: Competition and Class of Product

The four general types of price record exhibited by minor metals between 1945 and 1962—together with information about the geology of their occurrence and the technology of their extraction—provide a basis for some tentative generalizations about competitive conditions in the production of minor metals in the United States. Those metals and sources with prices that fluctuate daily or weekly come closest to being sold in purely competitive markets. Those metals with steadily rising prices come closest to having monopolistic controls over production. Metals with humped behavior or with falling prices lie between these extremes. As a rule, those exhibiting a humped price record have had raw material supply limitations which (under conditions of increased demand) overshadowed technologic advances. Those that exhibit decreasing prices have had technologic advances which (even under conditions of increased demand) overshadowed raw material supply limitations.

The generalizations about competition are related to the classification of minor metals developed early in the study, for both derive from the supply conditions for minor metals. Most metallurgical byproducts of major metals fall into the group for which raw material restrictions have overshadowed technologic advances. Those that do not fit this pattern are small-tonnage byproducts for which demand has never approached a short-run source material barrier. The same has been true for metallurgical byproducts and coproducts of other minor metals; metals of this class, together with those of the milling byproducts and coproducts class, are commodities for which technological advance has more than made up for any raw material deficiencies. No general conclusions of this sort can be stated for individually mined minor metals. A number, such as mercury and tungsten, are produced domestically under conditions approaching pure competition; sizable deposits of others are so rare that large profits can be earned. Finally, those metals for which there is evidence of monopoly power cut across class lines: in some cases, as with bismuth and molybdenum, the raw material barrier is important; in others, as with beryllium and vanadium, other factors are critical.

There can be no question about the technologic efficiency of the producers of minor metals in the United States. In case after case they have demonstrated their ability to extract minor metals from waste products or refractory ores, to separate metals from one another, and to reduce them to primary forms using a variety of techniques both old and new. For the most part, the production of minor metals is economically efficient as well. The lack of competition that was found for some metals in the years between 1945 and 1965 is neither important enough to alter the general picture nor strong enough to offset the interproduct competition that is so much a part of minor metal consumption.

APPENDICES

APPENDIX A

In many tables and at points in the text it is convenient to use chemical symbols rather than the names of the elements. The following list presents the symbols so used in alphabetical order:

As	arsenic	O	oxygen
Al	aluminum		
		Pb	lead
Be	beryllium	Pd	palladium
Bi	bismuth	Pt	platinum
Ca	calcium	Ra	radium
Cb	columbium	Rb	rubidium
Cd	cadmium	RE	rare-earth metals
Co	cobalt	Re	rhenium
Cs	cesium		
Cu	copper	Sb	antimony
		Sc	scandium
Fe	iron	Se	selenium
Ga	gallium	Ta	tantalum
Ge	germanium	Te	tellurium
		Th	thorium
Hf	hafnium	Ti	titanium
Hg	mercury	Tl	thallium
In	indium	V	vanadium
La	lanthanum	W	tungsten
Li	lithium		
		Y	yttrium
Mg	magnesium		
Mo	molybdenum	Zn	zinc
		Zr	zirconium
Na	sodium		

APPENDIX B

ABBREVIATIONS FOR CORPORATION NAMES

Many business concerns are commonly referred to by abbreviated names or initials, and this practice is followed in the text. The list below gives the official names for cases in which confusion might arise.

Alcoa — Aluminum Company of America
Amax — American Metal Climax, Inc.
American Potash — American Potash and Chemical Corporation
Asarco — American Smelting and Refining Company
Berylco — The Beryllium Corporation
Cominco — Consolidated Mining and Smelting Company of Canada, Ltd.
ICI — Imperial Chemical Industries, Ltd.
Inco — The International Nickel Company (of Canada), Ltd.
Lithium Corp. — Lithium Corporation of America
Mincon — Mineral Concentrates and Chemical Company, Inc.
Moly Corp. — Molybdenum Corporation of America
National Distillers — National Distillers and Chemical Corporation
Nuclear Corp. — Nuclear Corporation of America
PPG — Pittsburgh Plate Glass Company
Research Chemicals — Research Chemicals Division of Nuclear Corp.
SGMH — Société Générale Métallurgique de Hoboken
TMCA — Titanium Metals Corporation of America
UCC — Union Carbide Corporation
UCM — Union Carbide Metals Division of UCC (formerly Electromet)
United R & S — United Refining and Smelting Company
USI — U.S. Industrial Chemicals (division of National Distillers)
USSR&M — United States Smelting, Refining and Mining Company
VCA — Vanadium Corporation of America

APPENDIX C

This appendix brings together statistics on world production, domestic production, domestic consumption, exports, and imports of minor metal source materials and of primary products. Unfortunately, the deficiencies in the information are great. In addition to absence of relevant data and inconsistency from metal to metal in the data that are available, there are the problems of incomplete reporting and the withholding of data from publication because of the small number of producers.

The first section of this Appendix is a brief discussion of the framework used for presenting data on minor metals. The second section is comprised of the statistical tables themselves. Appendix Tables 1a and 1b present production and consumption data for minor metal sources and primary products respectively. These form the background data on which summary figures and tables in the text are based. Appendix Tables 2a and 2b give domestic production figures for selected minor metals sources and primary products, respectively, at five-year intervals since 1945. These figures form the basis for calculating the index numbers discussed in Chapter 5 and Appendix D. The data are all in physical rather than value terms. Not only is the availability of value data very spotty, but it is almost impossible to fill gaps in the data with reasonable estimates because it is so difficult to control or check the estimates.

Framework for the Data

Sources of Information

The sources for most of the figures in the appendix tables are issues of *Minerals Yearbook* and the *Materials Survey* series of the U.S. Bureau of Mines. Data given in any particular year of *Minerals Yearbook* are commonly revised in later years. In all cases, later data were accepted as superseding earlier data. The most recently available *Minerals Yearbook* was that for 1962, by which time most revisions in the figures for 1960 had been made. Therefore, statistics are presented through 1960; that year was also a convenient base for the index series. Trade journal articles and the American Bureau of Metal Statistics *Yearbook* were useful as supplementary sources.

Units and Reference of the Figures

Almost all figures are tabulated in units of thousands of pounds. (This requires that some metals, such as mercury, be tabulated in units different from those used commercially.) The only exceptions are commodities produced in very small amounts, data for which are tabulated in pounds. With the exception of a few world production figures, the distinction between metal contained in source materials and metal contained in primary products has been carefully observed. In the past, failure to make this distinction has led to a great deal of misunderstanding. (Exceptions must be made for the world production figures reported for certain metals, such as platinum, which are only available as a single combined total that incorporates both mine production and primary metal production data.) In addition, as much as possible, data refer to metal content rather than to gross weight. This is not a serious problem for primary products. Most data on primary products are given either in terms of metal content or in a form that permits calculation of metal content.[1] However, data on source materials are commonly reported in terms of gross weight. Estimates of content are possible only if standards have been established with respect to the grades of ore marketed or if the content of the ore is nearly constant.[2] Published world production totals for a few metals (such as lithium) add content data for some countries to recoverable metal or gross weight data for others; such data have been revised for use in these tables by recalculating all figures to the same basis.

There are a few variations on the basic production and consumption data. Some mine production figures are labeled "recoverable content." Recoverable content gives the amount of metal in an ore that can be expected to appear in primary products, allowing for smelting and refining losses. It is a useful statistic whenever mine production data must substitute for data on primary products. Similarly, data for shipments from mines are sometimes used in place of mine production data because the latter were not available or because only the former (which are actually recorded at mills where ores are assayed) were given in terms of content. Shipments data are more sensitive to the business cycle and do not necessarily reflect production in any one year. Finally, consumption figures were not available for some metals, but "apparent consumption" data could be found. Apparent consumption statistics are usually calculated according to the following formula: shipments plus imports minus exports minus changes in stocks. However, the

---

[1] Calculation of the metal content of chemical compounds by theoretical formulas (described in any elementary chemistry text) basically involves dividing the atomic weight of the metal in question by the total atomic weights of all elements in the compound, and then multiplying this ratio by the gross weight of compound to get the contained weight of that metal.

[2] Beryl, for example, is commercially traded on the basis of a content of 10 to 11 per cent BeO, and calculations of the Be content can be made on this basis. In other cases the technique described in footnote 1 can be used, since minerals are chemical compounds with definite compositional limits. But this procedure neglects impurities not only in the metal content of the mineral but also in the mineral content of the concentrate.

formula can be varied in a number of ways. It gives a more or less accurate estimate of actual consumption depending upon the number of factors taken into account. (In rare cases actual consumption data were available and an "apparent production" figure was calculated using an algebraic transformation of the above formula.)

Data Collected

The domestic data collected fall into two categories of production and two categories of consumption according to the analysis of production stages: mine production of sources for minor metals, consumption of mined products, production of primary minor metal products, and consumption of primary minor metal products. The first category is the output of the mining stage of production. The second and third are, respectively, input and output at the metallurgical reduction and refining stage. And the last is consumption of minor metals by their incorporation into producer and consumer goods.

The four sets of statistics (even if all were available) could be nothing more than an extended input-output series distinguished only by time lags, by stock and inventory adjustments, and by incomplete recovery of metal. However, in almost all cases one or more of the following intervene between them: imports or exports of source materials or of primary products; consumption of source material for purposes other than production of primary products; and minor metals that went unreported in import statistics because they were contained in major metal ores or alloys. Where any of the mined material is consumed as a mineral, there will be an inconsistency in the figures between the first and the second set if the consumption of mined products data include only that amount consumed for the purpose of making minor metal products; or there will be an inconsistency between the second and third set if the consumption of mined products data include the total consumed. Finally, for byproduct minor metals there will be inconsistency between data on primary products and that on mine production unless the latter include an estimate of the amount of minor metal contained in major metal ores. Many ratios shown as "not applicable" in the table result from such inconsistencies.

As a rule, it is not possible to obtain statistical information for all four categories. In particular, data are usually collected for consumption of mined products or for production of primary products, depending on whether it is easier to report input or output data. Also, world production statistics usually refer to mine production or to primary metal production, and can only be compared with the one appropriate domestic production figure. Even some domestic production statistics are not published for one reason or another. However, trade journal articles frequently report data that the Bureau of Mines is prohibited from disclosing. And in other cases it is possible to make estimates that fill in gaps in the published statistics.

Estimation

In some cases it was possible to estimate a figure that was either unavailable or unavailable in appropriate form. The intent of estimation has been to approach the form in which the data have been molded for this analysis. For example, there is a very accurate measure of the beryl mined in the United States since the Korean War because the federal government has purchased most of this beryl. But these data are in terms of gross weight, not in terms of beryllium content. Therefore, beryllium content was estimated from the gross weight figures, even though the result may be a less accurate measure of content than the original statistics are of gross weight. However, if an available set of statistics differs only by a small factor from the series defined here, the difference is usually ignored. For example, in recent years the Bureau of Mines figure for cadmium production includes only cadmium metal and neglects the small amount of cadmium produced as compounds directly from ore (less than 5 per cent of annual production). Nevertheless, no attempt has been made to correct the more recent figures. In other words, estimates have been made to get data in the correct statistical form, but not to make minor corrections.

Statistical Tables

General Notes for All Appendix Tables

The figures in the appendix tables refer either to source materials for minor metals or to primary minor metal products. Unless otherwise indicated those referring to source materials include all sources that can be identified, not just ore sources. Figures for the consumption of mined products include the total consumed and not just the amount consumed in the production of primary minor metal products, unless otherwise indicated. However, figures that refer to primary products include only commodities listed in col. 2 of Table 3, again unless otherwise indicated. For example, figures for titanium metal products include only titanium sponge metal. Columns for alloy production are therefore marked with the symbol for not applicable. Similarly, since rutile (the source material) is used for products other than sponge metal, division of sponge production by total rutile consumption would give an erroneous impression of recovery, and such a column would also have to be marked with the symbol for not applicable.

Most of the figures in the appendix tables refer to metal content, rather than to gross weight. Except when the data are in the form of ratios (which can be just as accurate when based on gross figures as when based on content figures), gross weight data are enclosed within parentheses. A gap in the table indicates that the figure is not available and has not been estimated. An asterisk is used to indicate an estimate by the author only where there is a chance of substantial error, and not where the desired figure had been calculated directly from U.S. Bureau of Mines figures.

Table 1a. U.S. Production and Consumption of Sources of Minor Metals, 1960

(Thousand pounds of metal content unless otherwise specified)

Minor metal	U.S. Mine production		Imports for consumption	Exports (excl. re-exports)	U.S. Consumption of mined products		World production
	Ore only	Est. all source material			Other than primary products in text Table 3	Total	
	(1)	(2)	(3)	(4)	(5)	(6)	(7)
Antimony	1,270	3,000*	12,942	0	0		122,000
Arsenic[1]	0	3,100*	large	0	0		
Beryllium	[2] 29	[2] 29	644*	1	small	698*	670*
Bismuth	0		large	0	0		
Cadmium	0		[3] 2,803	0	0		
Calcium				—not applicable[4]—			
Cesium and rubidium oxide	0	0	(7½)*	0	large		
Cobalt	[5] 1,570	[5] 1,570	314	0	0	2,062	[5] 33,400†
Columbium	0	0	(5,022)	(155)	negl.	1,538	[6] (7,020)†
Gallium	0			0	0		
Germanium	0	28*		0	0		
Hafnium			—(incl. with zirconium in zircon)—				
Indium	0				0		
Lithium	[7] (82,000)*	[7] (82,000)*	(102,198)	0	small	(184,200)*‡	(277,500)*†
Mercury			—(almost identical to mercury metal production—				
Molybdenum	68,237	68,237	0	23,341	small	44,784	89,500
Platinum-group totals	1.6		2.1	negl.	0		([8])
Radium	0	0	0	0	0	0	
Rare-earth oxides[1]	[2] (2,286)	[2] (2,286)	large	small	negl.	[9,10] 1,200	
Rhenium	0	4*	0	0	0		
Rubidium			—(incl. with cesium)—				
Scandium	0	0		0	0		
Selenium	0			0	0		
Sodium			—not applicable[4]—				
Tantalum	0	0	(710)	(32)	negl.	578	([6])
Tellurium	0			0	0		
Thallium	0			0	0		
Thorium[1]	[2] (2,286)	[2] (2,286)	large	negl.	large	[10] 62	
Titanium	[2] 10,900	[2] 10,900	32,950*	negl.	[11] 18,310	27,507	129,000*†
Tungsten	6,669	6,669	3,525	655*	small	11,605	66,049
Vanadium	[2] 16,094		5.5	large	0	17,600	[12] 20,332†
Yttrium			—(incl. with rare-earth metals)—				
Zirconium	(93,400)*	(93,400)*	(68,560)	(2,764)	[13] (158,400)‡	(178,000)‡	(377,000)†

NOTE: The source of most information was the "Minerals Yearbook," but other references were drawn upon more widely for this table than for others. In compiling the table and in making calculations very small quantities, such as production of laboratory amounts, have often been neglected. Negl. (negligible) is used in lieu of data with the following meaning: a quantity which probably amounts to less than 1 per cent of domestic consumption of the given metal. "Small" and "large" are used similarly for amounts between 1 per cent and 10 per cent and for amounts over 10 per cent, respectively.

( ) Figures in parentheses are gross weight.

* Author's estimate.

† World production excludes the Soviet Union and/or other non-Western nations that are significant producers.

‡ Apparent consumption.

[1] 1959 data.

[2] Mine shipments.

[3] Represents only that amount imported in flue dust and not that imported in zinc concentrates.

[4] Less than 1 per cent of source material used for production of metal.

[5] Recoverable content of ore and concentrate.

[6] Columbium plus tantalum totals shown in columbium row.

[7] Lithium content estimated at 2,460,000 pounds.

[8] Placer platinum ore included with refinery production in Appendix Table 1b.

[9] Total oxides including yttrium; combined rare-earth metals content by weight can be estimated at about 85 per cent of oxide content.

[10] Consumption from domestic sources only (imports classified).

[11] 9,197,000 pounds consumed for purpose of making titanium metal.

[12] Adjusted from recoverable metal to content of ore and concentrate.

[13] 19,600,000 pounds consumed for purpose of making zirconium metal and alloys.

Table 1b. U.S. Production and Consumption of Primary Minor Metal Products, 1960

(Thousand pounds of metal content unless otherwise specified)

Minor metal	U.S. Production				Imports for consumption (5)	Exports (excl. re-exports) (6)	New industrial consumption (7)	Secondary metal production (8)	World production (9)
	Metal (1)	Alloys (2)	Compounds (3)	Total (4)					
Antimony	7,330	1,312	11,266	19,908	15,946	0	26,542	40,208	112,200*
Arsenic[1]	negl.	0	7,860		29,365	0	40,330‡	0	71,200†
Beryllium (metal equivalents)	[2]200	[2]158*	[2]>47*	[2]405*	<1	4*		intraplant	
Bismuth	1,400*	small	0		1,167	95*	1,527	large	5,200
Cadmium	10,180	0	small		942	2,448	10,127	small	24,800
Calcium metal	n.ap.	n.ap.	n.ap.		12	0			100*
Cesium and rubidium (lb.)	175	0	(5,922)		0	0		0	
Cobalt	260	[4](1,384)	small	1,650*	11,856	0	8,690	[3]240	See Table 1a
Columbium (metal only except for cols. 2 and 3)				n.ap.	[4]5	small	200	large	
Gallium				0.2–0.5*					
Germanium	54	0	[5]n.ap.		17*	0	54	intraplant	120*†
Hafnium	[6]	0	107						
Indium	6–10*	0	small		30–35*	0	40*		
Lithium (in carbonate equivalents)	n.ap.	n.ap.	1,500*		small	small	1,500		2,200*
Mercury	2,525	0	0		1,481	27	3,889	407	18,392
Molybdenum (in trioxide)	n.ap.	[7]n.ap.	40,337		24	6,586	31,837		80,000*
Platinum-group totals	3.5	negl.	negl.		44.3	4.5	53.2	5.3	[8]87.6
Platinum	2.4	negl.	negl.		16.3	3.4	22.3	2.7	
Palladium	0.7	negl.	negl.		25.2	}1.1{	28.4	2.4	
Other platinum-group metals	0.4	negl.	negl.		2.8		2.5	0.2	
Radium	0	0	0	0	<0.1	negl.	<0.1		

Material	1	2	3	4	5	6	7	8
Rare-earth oxides[1,9]	n.ap.							
Rhenium		n.ap.		1.0	(17)	<1.0		
Rubidium (included with cesium)		n.ap.	<100	0	(15)	3,000‡		2
Scandium (lb.)	n.ap.	n.ap.	small	2			60*	
Selenium	539	n.ap.	small	162		714‡		1,671†
Sodium metal	228,000	n.ap.	n.ap.					
Tantalum (metal only except for cols. 2 and 3)	300	(4)	small		>1	400		large
Tellurium	271	n.ap.	small	15		243‡	large	389†
Thallium		0	2–5				negl.	
Thorium (non-energy uses only)	10,622	n.ap.	n.ap.			168		
Titanium (sponge metal only)				4,462	1,758	10,974	4,054[3]	18,400
Tungsten			9,397	363	>166	8,535	561[3]	
Vanadium (in pentoxide)	n.ap.	n.ap.	10,990	negl.	large	4,480	large	
Yttrium				1*	<996	<1		
Zirconium (sponge metal only)	2,846	n.ap.	n.ap.			1,529*	400	16,000*

NOTE: See Appendix Table 1a.

( ) Figures in parentheses are gross weight.

* Author's estimate.

† World production excludes the Soviet Union and/or other non-Western nations that are significant producers.

‡ Apparent consumption.

n.a.p.—not applicable.

[1] 1959 data.

[2] Shipments rather than production.

[3] Data for secondary metal represent consumption rather than production.

[4] Columbium plus tantalum totals shown in columbium row.

[5] All germanium dioxide produced is used to make metal.

[6] 70,000 pounds of sponge hafnium produced from the hafnia shown in column (3).

[7] 2,602,000 pounds of molybdenum metal powder produced from oxide.

[8] Includes placer platinum as well as refinery production.

[9] Total oxides; combined rare-earth metals content by weight can be estimated at about 85 per cent of oxide content.

Table 2a. U.S. Production of Sources for Selected Minor Metals at Five-Year Intervals, 1945–60

Metal	U.S. production (thousand pounds)				Price weight used for computing index ($/lb.)[1]
	1945	1950	1955	1960	
Antimony (ore and concentrate only)	3,860	4,994	1,266	1,270	0.12
Beryllium	2.8	40	40	29	6.39
Cobalt	1,100	809	2,609	1,570	1.56 × 0.78
Columbium-tantalum (ore and concentrate only)	(6.6)	(1.0)	(13)	0	(2)
Lithium	250	680	5,400*	2,460*	3.564 × 0.595
Molybdenum	30,802	28,480	61,781	68,237	1.25
Platinum-group	2.1	2.6	1.6	1.6	[3]914.00
Rare-earth metals and thorium	0	170	1,020	935	.30
Titanium	7,700	7,450	10,350	10,900	.081
Tungsten	5,389	3,965	15,833	6,669	1.39
Vanadium (recoverable Vanadium)	2,371*	3,197	6,572	9,942	2.464
Zirconium	(5,362)	(28,000)*	(56,220)	(93,400)*	49.00 × 0.49

* Author's estimate.

( ) Figures in parentheses are gross weight.

[1] Some metals have two weighting factors. In such cases the first is a price weight and the second is a recovery factor. The double weighting was necessary if no price was available for the source material itself so that a price for metal had to be substituted.

[2] Domestic price unavailable. U.S. Bureau of Mines value of production data used directly. This somewhat underweights columbium-tantalum concentrates in 1945 and 1950, and overweights them in 1955.

[3] The price weight for platinum ore assumes that the crude platinum is two-thirds platinum and one-third palladium.

SOURCE: U.S. Bureau of Mines, "Minerals Yearbook."

Table 2b.   U.S. Production of Selected Primary Minor Metal Products at Five-Year Intervals, 1945–60

Metal	U.S. production (thousand pounds)				Price weight used for computing index ($/lb.)
	1945	1950	1955	1960	
Antimony	45,984	24,642	20,402	19,908	0.291
Arsenic	36,864	20,095	16,321	9,463*	0.528
Bismuth	1,456*	2,100*	1,232*	1,400*	2.25
Cadmium	8,384	9,190	9,754	10,180	1.23
Cobalt[1]	3,750*	1,975*	3,800*	1,600*	1.56
Germanium	<1	3*	39*	54	136.00
Lithium[1]	300*	456*	1,640*	1,500*	3.564
Mercury	2,338	345	1,441	2,525	2.773
Molybdenum	32,406	25,348	37,774	40,337	1.46
Platinum metal	11.1	3.9	3.6	2.4	1,196.00
Palladium metal	2.0	0.8	0.4	0.7	350.00
Rare-earth metals[1]	560	2,800*	3,050	3,050	0.30
Selenium	459	511	699	539	6.75
Tellurium	33	107	180	271	3.50
Thorium[1]	0	46	65	218	12.50
Titanium	0	150	14,796	10,622	1.60
Vanadium	3,500*	3,640	7,338	10,990	2.464
Zirconium	0	5*	374	2,846	6.25

* Author's estimate.

[1] Production of primary products for these metals has been estimated from consumption. This is possible only when imports and exports of metal products are small. The data for cobalt are based on domestic consumption of cobalt ore and concentrates, and are adjusted for a recovery of 78 per cent. The data for thorium are based on data for consumption of thoria in alloys with magnesium. The data for rare-earth metals are based on consumption or apparent consumption of rare-earth oxides and are adjusted for an assumed 85 per cent content of metal. The data for lithium are based on estimates of equivalent lithium carbonate consumption.

SOURCE: U.S. Bureau of Mines, "Minerals Yearbook."

APPENDIX D

PREPARATION OF INDEX SERIES

In order to study changes in the production of minor metals as a whole, it was necessary to combine the statistical information about each metal into a composite set of figures representative of all of them. The admonitions that must accompany such index numbers are well known.[1] Here it remains to discuss their computation.

The two main problems in the computation of the index numbers were selection of the unit prices to be used in weighting and selection of a weighting formula. The weighting formula will be discussed first, and for this purpose it can be assumed that the price problem has been solved. The simplest of the many possible weighting formulas was used. The commodities were weighted by their unit prices in 1960, the final year of the series. In view of the approximations made in data collection, it did not seem worthwhile to use a more complex type of weighting. Different prices were, of course, required for the different stages of production. The year 1960 was a useful base year because prices for many minor metals were stable over the period 1957 through 1961. And by 1960, all of the newer metals included in the index were well beyond the stage of laboratory production and of attendant noncommercial prices.

The first mathematical operation was to multiply each quantity statistic q for each metal by an appropriate price p. This operation formed series showing for year t what the value of production for commodity n (produced at that stage in year t) would have been had it been produced at 1960 prices. The second operation was to sum the weighted figures for each year over all the metals for which data was available, thus yielding index numbers of the form:

$$\sum^{n} q_t \, p_{1960} .$$

Then, for ready comparison with the Federal Reserve Board index and the Bureau of Mines index, both of which have used the 1947–49 average as a base for recalculation to 100, the index numbers for 1945 and 1950 were

[1] See, for example, Frederick C. Mills, *Statistical Methods* (revised edition; New York: Henry Holt and Co., 1938), pp. 161–67, 180–210, and 306–310.

averaged, and this average was used as the basis for recalculation to 100 of the individual index numbers. The final index formula is:

$$\frac{2 \left(\sum_{}^{n} q_t \, p_{1960} \right)}{\sum_{}^{n} q_{1945} \, p_{1960} + \sum_{}^{n} q_{1950} \, p_{1960}} .$$

Note that the summation sign refers only to different metals, and thus the formula yields a weighted index number at each year and at each stage of production.

The selection of price weights to use in the formula was guided by the form of the commodity. That is, production of source material was weighted by the price of metal in ore, and production of a primary product was weighted by the price of that primary product. The actual prices used in weighting are shown in Appendix Tables 2a and 2b.[2] Inasmuch as these prices are value per pound of metal content, they will not agree with prices quoted on a different basis. The following rules were used in selecting individual price weights:

(1) If average unit value figures were available or could be calculated from total value, they were used as weights.

(2) If value figures were unavailable, quoted prices (f.o.b. producer's plant, if possible) were used as weights. If a price range was quoted, the lower limit was utilized. (The price for large lots has been more responsive to technologic advances than has that for small lots. Moreover, in 1960 there was some discounting from quoted prices.) If the quoted price changed during 1960, an average quoted price was calculated. For example, if some metal was quoted at $4 per pound for eight months and $7 per pound for four months, the price weight would be taken as $5.

[2] Use of 1960 physical data with these price weights gives the two estimates of value of production for 1960 that were quoted in Chapter 5.

INDEX

141

For Product Safety Concerns and Information please contact our
EU representative GPSR@taylorandfrancis.com Taylor & Francis
Verlag GmbH, Kaufingerstraße 24, 80331 München, Germany